Inverse Spectral Theory

Jürgen Pöschel
Universität Bonn
Mathematisches Institut
D-5300 Bonn
Federal Republic of Germany

Eugene Trubowitz
Mathematik
ETH-Zentrum
CH-8092 Zürich
Switzerland

ACADEMIC PRESS, INC.
Harcourt Brace Jovanovich, Publishers
Boston Orlando San Diego
New York Austin London Sydney
Tokyo Toronto

ACADEMIC PRESS, INC.
Orlando, Florida 32887

United Kingdom Edition published by
ACADEMIC PRESS INC. (LONDON) LTD.
24–28 Oval Road, London NW1 7DX

1003134474

Library of Congress Cataloging-in-Publication Data

Pöschel, Jürgen.
  Inverse spectral theory.

  (Pure and applied mathematics ;   )
  Bibliography: p.
  Includes index.
  1. Spectral theory (Mathematics) I. Trubowitz,
Eugene. II. Title. III. Series: Pure and applied
mathematics (Academic Press) ;   .
  QA3.P8   510 s [515.7'222]   86-47801
  [QA320]
  ISBN 0-12-563040-9 (alk. paper)

87 88 89 90   9 8 7 6 5 4 3 2 1
Printed in the United States of America

Contents

Preface

This book is based on lectures given during the winter semester of 1980 at the Courant Institute of Mathematical Sciences in New York and during the winter semester of 1981 at the Eidgenössische Technische Hochschule in Zürich. The present work, the result of a collaborative effort, incorporates many extensions and improvements. We have made a strenuous attempt to keep the discussion self-contained and as simple as possible to make it easily accessible to a general mathematical audience. On the other hand, the approach is novel, some of the material is new, and the book may therefore interest the expert reader.

In 1836 both Sturm and Liouville ([St], [Li]) published articles in the same volume of the Journal de Mathématique concerning boundary value problems for the differential equation

$$(1) \qquad -y'' + q(x)y = \lambda y, \qquad 0 \le x \le 1.$$

Here, λ is a complex parameter and q is a real-valued function which we shall assume is square-integrable over the unit interval $[0, 1]$. Sturm and Liouville asked whether there exist nontrivial solutions of equation (1) satisfying boundary conditions of the form

$$(2) \qquad \begin{aligned} y(0) \cos \alpha + y'(1) \sin \alpha &= 0 \\ y(1) \cos \beta + y'(1) \sin \beta &= 0 \end{aligned}$$

where α, β are real numbers between 0 and π.

A complex number λ is called an eigenvalue of q and α, β if the boundary value problem (1) and (2) can be solved. The corresponding nontrivial solutions are called eigenfunctions of q and α, β for λ. The collection of all eigenvalues is the spectrum of the boundary value problem.

As is well known, a comprehensive spectral theory of general ordinary and partial differential operators has been developed that has its roots in the simple example discussed above. These sophisticated techniques give us insight into the nature of spectra and the behavior of eigenfunctions for a wide variety of problems.

At this time, there is no general inverse spectral theory—even the simplest cases require considerable ingenuity for their resolution. In the present context let us consider two specific questions.

Fix α and β. First we ask to what extent is the coefficient function q determined by the spectrum of the associated boundary value problem? Secondly, is it possible to characterize all sets of numbers that arise as the spectrum of some q for the fixed boundary conditions corresponding to α, β? It may come as a mild surprise that both questions have complete answers.

The boundary conditions (2) are self-adjoint in the usual sense that for any pair of functions satisfying them

$$\langle Qf, g \rangle = \langle f, Qg \rangle$$

where $Q = -(d^2/dx^2) + q(x)$ and $\langle f, g \rangle = \int_0^1 f(x)\bar{g}(x)\, dx$. There is another class of self-adjoint boundary conditions for (1). Namely,

$$(3) \qquad \begin{pmatrix} a & b \\ c & d \end{pmatrix}\begin{pmatrix} y(1) \\ y'(1) \end{pmatrix} = e^{ik}\begin{pmatrix} y(0) \\ y'(0) \end{pmatrix}$$

where a, b, c, d are real with $ad - bc = 1$ and k is an arbitrary real number. In fact, every self-adjoint boundary condition for (1) is of the form (2) or (3). The same questions can be posed for the boundary conditions (3). They can also be answered.

None of these inverse problems can really be posed unless a class of coefficients q is specified in advance. We take $L^2 = L_R^2[0, 1]$, the Hilbert space of all real-valued square-integrable functions on $[0, 1]$. It is possible to work with other spaces of functions, or even measures, but the basic ideas and constructions are clearest for L^2, as with Fourier series. For example, geometry is simpler in a Hilbert space than in a general Banach space: the gradient of a differentiable function $f(q)$ on L^2 is defined, and therefore the normal field to the level set $f(q) = c$ is easily visualized—this picture will be important for us in Chapter 4.

It is possible to systematically answer the questions raised above for all these different boundary conditions (see, for example, [IT], [IMT], and [MT]), but it is a large task. Our intentions for this book are by contrast rather modest. We shall discuss a single example—the Dirichlet problem of finding solutions to (1) satisfying the Dirichlet boundary conditions

(4) $y(0) = 0, \qquad y(1) = 0.$

This apparently naive case already exhibits many interesting features and displays the basic techniques in a transparent form. It is a short story of its own.

There is a simple physical interpretation of the inverse Dirichlet problem. The displacement $u = u(x, t)$, $0 \leq x \leq L$, of a freely vibrating inhomogeneous stiring of length L and variable mass density $\rho(x) > 0$ satisfies the wave equation

$$\rho(x)u_{tt} = u_{xx}$$

and the boundary conditions

$$u(0, t) = 0, \qquad u(L, t) = 0.$$

A periodic vibration of the form

$$u = y(x)(a \cos \omega t + b \sin \omega t),$$

with frequency ω, is called a pure tone. Separating variables, y must satisfy

(5) $y'' + \omega^2 \rho(x)y = 0$

and

(6) $y(0) = 0, \qquad y(L) = 0.$

From this point of view it is reasonable to ask: how many ways can mass be distributed along a string to produce a given set of frequencies, and how can one tell when a set of numbers is the set of frequencies of an actual vibrating string? Liouville made the observation (see, for example, [MW], p. 51) that (5) and (6) can be transformed into (1) and (4) when ρ is twice continuously differentiable. Hence, the inverse Dirichlet problem can be interpreted as posing natural questions about the pure tones of strings.

The necessary facts about the Dirichlet boundary value problem are derived in the first two chapters. Most of them are well known. However, a few results may be unfamiliar since we emphasize that various quantities, such as fundamental solutions, eigenvalues and eigenfunctions, are themselves analytic functions of the coefficient q. This perspective is essential for us.

Beginning with Chapter 3, the inverse Dirichlet problem is formulated and in due course completely solved. First, we answer the question: to what extent is a coefficient function determined by its Dirichlet spectrum? This is done by describing the set of all coefficients with the same, given, Dirichlet spectrum, the so-called isospectral set. Isospectral sets turn out to be infinite dimensional manifolds with many special properties.

In the second part we address the problem of characterizing all sequences of Dirichlet eigenvalues. To make this problem more tangible, imagine shifting one of the eigenvalues in the spectrum of a function p to the right, or the left, by a small amount. Is the new sequence still realizable as the Dirichlet spectrum for some other coefficient q, or does it violate some internal constraint that is satisfied by an actual sequence of eigenvalues? A full characterization is given in Chapter 6.

It is necessary to point out that our treatment of this part of inverse spectral theory is very personal and consequently does not depend on or develop other established approaches and contains few references to published material. We refer to [CS] for an extensive bibliography and historical information, and to Borg [Bo], a student of Beurling, who wrote the first important paper on the subject, as well as to Gelfand and Levitan [GL] who were the first to discover a method for constructing coefficients with a given spectral measure.

We recommend that the reader wishing to get a quick overall impression read the first few pages of each chapter and then skim the statements of the theorems.

In this volume we have presented our subject in isolation and have not explored several important connections with other areas of mathematics. Were we to continue and write a second volume we would likely discuss, among other topics, the periodic problem and the relationships to classical mechanics, algebraic geometry, and the Korteweg de Vries equation of fluid dynamics.

Finally, it is a pleasure to thank P. Deift, H.P. McKean, T. Nanda, C. Tomei and especially E. Isaacson, J. Moser, and J. Ralston for their help. This book is dedicated to my father on the occasion of his seventy-fifth birthday.

Zürich, 1986   *E. Trubowitz*

1   A Fundamental Solution

The purpose of this chapter is to solve the initial value problem for the differential equation

(1) $$-y'' + q(x)y = \lambda y, \qquad 0 \le x \le 1.$$

Here, $\lambda \in \mathbb{C}$, the complex numbers, and $q \in L_{\mathbb{C}}^2 = L_{\mathbb{C}}^2[0, 1]$, the Hilbert space of all complex valued, square integrable functions on $[0, 1]$.[1]

The plan is to construct solutions $y_1(x, \lambda, q)$ and $y_2(x, \lambda, q)$ of equation (1) satisfying the initial conditions

$$y_1(0, \lambda, q) = y_2'(0, \lambda, q) = 1$$

$$y_1'(0, \lambda, q) = y_2(0, \lambda, q) = 0,$$

and to show that they are a fundamental solution. That is, any other solution y of equation (1) can be written as a linear combination of these two solutions, namely,

$$y(x) = y(0)y_1(x) + y'(0)y_2(x).$$

An analysis of this fundamental solution will then give us information about the behavior of all solutions of (1).

---

[1] Starting with Chapter 2, we restrict our attention to real valued functions q. However, it is important for us to develop the basic theory on the complex Hilbert space $L_{\mathbb{C}}^2$. See for example the proof of Theorem 3.1.

Starting with the next chapter, we are going to use y_1 and y_2 to study boundary value problems. It is a basic principle of the theory of linear ordinary differential equations that boundary value problems can be solved by first constructing appropriate solutions to the initial value problem. This principle is illustrated at the beginning of Chapter 2, where the Dirichlet problem for equation (1) is considered.

As the notation indicates, we regard y_1 and y_2 as functions of all three variables x, λ and q. This point of view is essential for the applications we have in mind. In fact, we are going to construct y_1 and y_2 by expanding them as power series in q. It will be shown that they are entire functions on $\mathbb{C} \times L_{\mathbb{C}}^2$. See Appendix A for the basic facts about analysis on Banach spaces, and in particular analytic maps.

Before we continue, we must explain what we mean by a solution of equation (1). After all, the coefficient q is in $L_{\mathbb{C}}^2$, so that the equation only makes sense almost everywhere. By definition, a function y is a *solution* of equation (1), if it is continuously differentiable, y' is absolutely continuous, and the equation holds almost everywhere.[2]

A common method for solving differential equations is to look for solutions in the form of a power series. We are going to apply this method to equation (1). To see what is involved, we first consider the simple problem of finding the solution of

$$-u'' = \lambda u, \qquad 0 \le x \le 1,$$

with the initial conditions

$$u(0) = 1, \qquad u'(0) = 0.$$

Notice that $u(x, \lambda) = y_1(x, \lambda, 0)$.

Suppose that u is given by a power series in λ:

$$u(x, \lambda) = \sum_{n \ge 0} u_n(x)\lambda^n.$$

---

[2] Recall that an absolutely continuous function u on $[0, 1]$ has a derivative u' almost everywhere, which is integrable and satisfies

$$u(x) = u(0) + \int_0^x u'(t)\, dt.$$

Conversely, if v is an integrable function on $[0, 1]$, then $u(x) = \int_0^x v(t)\, dt$ is absolutely continuous, and $u' = v$ almost everywhere.

The zeroth coefficient is the solution to $-u'' = \lambda u$ for $\lambda = 0$, satisfying $u_0(0) = 1$, $u_0'(0) = 0$. Thus,

$$u_0(x) = 1.$$

Formally, differentiating the power series two times with respect to x, using the differential equation and equating coefficients, we find that

$$-u_n'' = u_{n-1}, \qquad n \geq 1.$$

The initial conditions for u_n are

$$u_n(0) = 0, \qquad u_n'(0) = 0, \qquad n \geq 1,$$

since $u(0) = 1 + \sum_{n \geq 1} u_n(0) \lambda^n = 1$ and $u'(0) = \sum_{n \geq 1} u_n'(0) \lambda^n = 0$ for all λ. Integrating twice,

$$u_n(x) = -\int_0^x \int_0^t u_{n-1}(s) \, ds \, dt,$$

so that by induction

$$u_n(x) = \frac{(-1)^n}{(2n)!} x^{2n}, \qquad n \geq 1.$$

Consequently,

$$u(x, \lambda) = \sum_{n \geq 0} \frac{(-1)^n}{(2n)!} x^{2n} \lambda^n = \cos \sqrt{\lambda}\, x,$$

which is no surprise at all. The point of this calculation is that the coefficients u_n, $n \geq 1$, are determined iteratively by solving the inhomogeneous form of the differential equation $-u'' = \lambda u$ for $\lambda = 0$.

We are going to imitate the last paragraph and construct y_1 and y_2 as power series in q. Judging by the above, a necessary prerequisite is the solution of the inhomogeneous form of equation (1) for $q = 0$.

Lemma 1. *Let $f \in L_c^2$ and $a, b \in \mathbb{C}$. The unique solution of*

$$-y'' = \lambda y - f(x), \qquad 0 \leq x \leq 1$$

satisfying

$$y(0) = a, \qquad y'(0) = b$$

is

$$y(x) = a \cos \sqrt{\lambda}\, x + b \frac{\sin \sqrt{\lambda}\, x}{\sqrt{\lambda}} + \int_0^x \frac{\sin \sqrt{\lambda}\,(x - t)}{\sqrt{\lambda}} f(t)\, dt.$$

We will also use the notation

$$c_\lambda(x) = \cos \sqrt{\lambda}\, x, \qquad s_\lambda(x) = \frac{\sin \sqrt{\lambda}\, x}{\sqrt{\lambda}}.$$

Then the solution reads

$$y(x) = ac_\lambda(x) + bs_\lambda(x) + \int_0^x s_\lambda(x - t) f(t)\, dt.$$

Proof of Lemma 1.   Consider the integral

$$y_f(x) = \int_0^x s_\lambda(x - t) f(t)\, dt.$$

By the addition theorem for the sine function,

$$y_f(x) = s_\lambda(x) \int_0^x c_\lambda(t) f(t)\, dt - c_\lambda(x) \int_0^x s_\lambda(t) f(t)\, dt.$$

Since $c_\lambda f$ and $s_\lambda f$ are integrable, y_f is absolutely continuous. Hence,

$$y_f'(x) = c_\lambda(x) \int_0^x c_\lambda(t) f(t)\, dt + \lambda s_\lambda(x) \int_0^x s_\lambda(t) f(t)\, dt$$

almost everywhere, and consequently everywhere, because the right hand side is continuous in x.[3] It follows from another differentiation that y_f is a particular solution of $-y'' = \lambda y - f(x)$ with

$$y_f(0) = 0, \qquad y_f'(0) = 0.$$

Since c_λ and s_λ solve the homogeneous equation,

$$y(x) = ac_\lambda(x) + bs_\lambda(x) + y_f(x)$$

is a solution of $-y'' = \lambda y - f(x)$ satisfying the correct initial conditions.

---

[3] If u is absolutely continuous and v is continuous with $u' = v$ almost everywhere, then u is continuously differentiable, and $u' = v$ everywhere. For, $u(x) = u(0) + \int_0^x u'(t)\, dt = u(0) + \int_0^x v(t)\, dt$ and thus $u' = v$ everywhere by the fundamental theorem of calculus.

To prove uniqueness, suppose that $\tilde{y}$ is another solution of the inhomogeneous differential equation satisfying the same initial conditions. Then the difference $v = y - \tilde{y}$ satisfies

$$-v'' = \lambda v, \qquad v(0) = 0, \qquad v'(0) = 0.$$

It follows that v vanishes identically, hence $y = \tilde{y}$.   ∎

We are now ready to construct y_1 and y_2. We begin with y_1. Suppose that $y_1(x, \lambda, q)$ is given as a power series in q. That is,

$$y_1(x, \lambda, q) = C_0(x, \lambda) + \sum_{n \geq 1} C_n(x, \lambda, q),$$

where

$$C_n(x, \lambda, q) = C_n(x, \lambda, q_1, ..., q_n) \Big|_{q_1 = \cdots = q_n = q},$$

and $C_n(x, \lambda, q_1, ..., q_n)$ is a bounded, multi-linear symmetric form on $L_{\mathbb{C}}^2 \times \cdots \times L_{\mathbb{C}}^2$ for each x and λ. Thus, C_1 is linear in q, C_2 is quadratic in q, and so on.

The zeroth order term is determined by setting $q = 0$. This gives

$$C_0(x, \lambda) = \cos \sqrt{\lambda}\, x.$$

Proceeding as before, we formally differentiate the power series for y_1 two times with respect to x,[4] use the differential equation and equate terms, which are homogeneous of the same degree in q. We obtain the differential equation

$$-C_n'' = \lambda C_n - q C_{n-1}, \qquad n \geq 1.$$

The initial conditions are

$$C_n(0, \lambda, q) = 0, \qquad C_n'(0, \lambda, q) = 0, \qquad n \geq 1,$$

since $y_1(0) = 1 + \sum_{n \geq 1} C_n(0) = 1$ and $y_1'(0) = \sum_{n \geq 1} C_n'(0) = 0$ for all q.[5] It follows from Lemma 1 that

---

[4] Here, one must not differentiate with respect to q using the chain rule. q is a variable in its own right. Consider, for example, the function

$$F(x, q) = x \int_0^1 q(t)\, dt.$$

[5] Replacing q by tq, the sums become power series in t with coefficients $C_n(0, \lambda, q)$ and $C_n'(0, \lambda, q)$ respectively. These power series vanish identically in t, so all coefficients must vanish.

$$C_n(x, \lambda, q) = \int_0^x s_\lambda(x - t)q(t)C_{n-1}(t, \lambda, q)\,dt, \qquad n \geq 1.$$

Hence,

$$C_1(x, \lambda, q) = \int_0^x s_\lambda(x - t_1)q(t_1)C_0(t_1, \lambda)\,dt_1$$

$$= \int_0^x c_\lambda(t_1)s_\lambda(x - t_1)q(t_1)\,dt_1,$$

and

$$C_2(x, \lambda, q) = \int_0^x s_\lambda(x - t_2)q(t_2)C_1(t_2, \lambda, q)\,dt_2$$

$$= \int_0^x s_\lambda(x - t_2)q(t_2)\left(\int_0^{t_2} c_\lambda(t_1)s_\lambda(t_2 - t_1)q(t_1)\,dt_1\right)dt_2$$

$$= \int_{0 \leq t_1 \leq t_2 \leq t_3 = x} c_\lambda(t_1) \prod_{i=1}^{2}[s_\lambda(t_{i+1} - t_i)q(t_i)]\,dt_1\,dt_2.$$

Proceeding by induction,

$$(2) \quad C_n(x, \lambda, q) = \int_{0 \leq t_1 \leq \cdots \leq t_{n+1} = x} c_\lambda(t_1) \prod_{i=1}^{n}[s_\lambda(t_{i+1} - t_i)q(t_i)]\,dt_1 \cdots dt_n$$

for $n \geq 1$.

Thus, we have written

$$y_1(x, \lambda, q) = c_\lambda(x) + \sum_{n \geq 1} C_n(x, \lambda, q)$$

as a formal power series in q. Such expansions were already known to Hermann Weyl [We].

The same reasoning applies to $y_2(x, \lambda, q)$ and yields

$$y_2(x, \lambda, q) = s_\lambda(x) + \sum_{n \geq 1} S_n(x, \lambda, q),$$

where

$$(3) \quad S_n(x, \lambda, q) = \int_{0 \leq t_1 \leq \cdots \leq t_{n+1} = x} s_\lambda(t_1) \prod_{i=1}^{n}[s_\lambda(t_{i+1} - t_i)q(t_i)]\,dt_1 \cdots dt_n$$

for $n \geq 1$.

Problem 1.   Show that, for $n \geq 1$,

$$C_n(x, \lambda, q) = \int_{0 \leq t_1 \leq \cdots \leq t_{n+1} = x} \prod_{i=1}^{n} (t_{i+1} - t_i)(q(t_i) - \lambda) \, dt_1 \cdots dt_n$$

and

$$S_n(x, \lambda, q) = \int_{0 \leq t_1 \leq \cdots \leq t_{n+1} = x} t_i \prod_{i=1}^{n} (t_{i+1} - t_i)(q(t_i) - \lambda) \, dt_1 \cdots dt_n .$$

[Hint: Use that $y_j(x, \lambda, q) = y_j(x, 0, q - \lambda)$ for $j = 1, 2$.]

So far, these are just formal calculations. We now show that these series converge to genuine solutions of equation (1).

Below, the standard notations

$$\langle f, g \rangle = \int_0^1 f(t)\bar{g}(t) \, dt, \qquad \|f\| = \sqrt{\langle f, f \rangle}$$

are used for the inner product and norm on $L_{\mathbb{C}}^2$.

Theorem 1.   *The formal power series for $y_1(x, \lambda, q)$ and $y_2(x, \lambda, q)$, whose coefficients are given by* (2) *and* (3) *respectively, converge uniformly on bounded subsets of* $[0, 1] \times \mathbb{C} \times L_{\mathbb{C}}^2$ *to the unique solutions of*

$$-y'' + q(x)y = \lambda y, \qquad 0 \leq x \leq 1,$$

satisfying the initial conditions

$$y_1(0, \lambda, q) = y_2'(0, \lambda, q) = 1$$

$$y_1'(0, \lambda, q) = y_2(0, \lambda, q) = 0.$$

Moreover, they satisfy the integral equations

$$y_1(x, \lambda, q) = \cos \sqrt{\lambda} \, x + \int_0^x \frac{\sin \sqrt{\lambda} \, (x - t)}{\sqrt{\lambda}} q(t) y_1(t, \lambda, q) \, dt$$

$$y_2(x, \lambda, q) = \frac{\sin \sqrt{\lambda} \, x}{\sqrt{\lambda}} + \int_0^x \frac{\sin \sqrt{\lambda} \, (x - t)}{\sqrt{\lambda}} q(t) y_2(t, \lambda, q) \, dt,$$

and the estimate

$$|y_1(x, \lambda, q)|, \, |y_2(x, \lambda, q)| \leq \exp(|\operatorname{Im} \sqrt{\lambda}|x + \|q\| \sqrt{x}).$$

Proof.   The proof is given for y_1. It is the same for y_2.

Recall the elementary inequalities

$$|c_\lambda(x)| = \tfrac{1}{2}|e^{i\sqrt{\lambda}x} + e^{-i\sqrt{\lambda}x}| \leq \exp(|\text{Im }\sqrt{\lambda}|x)$$

and, for $0 \leq x \leq 1$,

$$|s_\lambda(x)| = \left| \int_0^x c_\lambda(t)\, dt \right| \leq \int_0^x \exp(|\text{Im }\sqrt{\lambda}|t)\, dt \leq \exp(|\text{Im }\sqrt{\lambda}|x).$$

The nth term in the expansion of y_1 is then majorized by

$$|C_n(x, \lambda, q)| \leq \int_{0 \leq t_1 \leq \cdots \leq t_{n+1} = x} |c_\lambda(t_1)| \prod_{i=1}^{n} |s_\lambda(t_{i+1} - t_i)q(t_i)|\, dt_1 \cdots dt_n$$

$$\leq \exp(|\text{Im }\sqrt{\lambda}|x) \int_{0 \leq t_1 \leq \cdots \leq t_{n+1} = x} \prod_{i=1}^{n} |q(t_i)|\, dt_1 \cdots dt_n.$$

The value of the integral in the last line does not change under permutation of $t_1, \ldots, t_n$. Moreover, the union of all the permuted regions of integration is $[0, x]^n$. It follows that

$$\int_{0 \leq t_1 \leq \cdots \leq t_n = x} \prod_{i=1}^{n} |q(t_i)|\, dt_1 \cdots dt_n = \frac{1}{n!} \int_{[0, x]^n} \prod_{i=1}^{n} |q(t_i)|\, dt_1 \cdots dt_n$$

$$= \frac{1}{n!} \left[\int_0^x |q(t)|\, dt \right]^n$$

$$\leq \frac{1}{n!} (\|q\| \sqrt{x})^n$$

by the Schwarz inequality. Thus, we obtain the estimate

$$|C_n(x, \lambda, q)| \leq \frac{1}{n!} \exp(|\text{Im }\sqrt{\lambda}|x)(\|q\| \sqrt{x})^n.$$

This shows that the formal power series for y_1 converges uniformly on bounded subsets of $[0, 1] \times \mathbb{C} \times L^2_{\mathbb{C}}$ to a continuous function. Summing the majorants we obtain the estimate for y_1.

Next, we observe that

$$y_1(x, \lambda, q) = c_\lambda(x) + \sum_{n \geq 1} C_n(x, \lambda, q)$$

$$= c_\lambda(x) + \sum_{n \geq 1} \int_0^x s_\lambda(x - t)q(t)C_{n-1}(x, \lambda, q) \, dt$$

$$= c_\lambda(x) + \int_0^x s_\lambda(x - t)q(t)\left(\sum_{n \geq 1} C_{n-1}(x, \lambda, q)\right) dt$$

$$= c_\lambda(x) + \int_0^x s_\lambda(x - t)q(t)y_1(x, \lambda, q) \, dt,$$

since the uniform convergence allows us to interchange summation and integration. This verifies the integral equation for y_1. It follows just as in the proof of Lemma 1 that y_1 is a solution of $-y'' = \lambda y - q(x)y$ satisfying the correct initial conditions.

To prove uniqueness, suppose that $\tilde{y}_1$ is another solution of equation (1) satisfying the same initial conditions as y_1. By Lemma 1,

$$\tilde{y}_1(x) = c_\lambda(x) + \int_0^x s_\lambda(x - t)q(t)\tilde{y}_1(t) \, dt.$$

The difference $v = y_1 - \tilde{y}_1$ then satisfies

$$v(x) = \int_0^x s_\lambda(x - t)q(t)v(t) \, dt,$$

hence

$$|v(x)|^2 \leq \int_0^x |s_\lambda(x - t)q(t)|^2 \, dt \cdot \int_0^x |v(t)|^2 \, dt$$

$$\leq c \int_0^x |v(t)|^2 \, dt$$

by the Schwarz inequality with $c = \|q\|^2 \max_{0 \leq t \leq 1}|s_\lambda^2(t)|$. It follows that the nonnegative function $e^{-cx}\int_0^x |v(t)|^2 \, dt$ has a nonpositive derivative on $[0, 1]$. Since it vanishes at zero, it must vanish identically on $[0, 1]$. Thus, $v = 0$, proving uniqueness. ∎

Certainly, Theorem 1 is familiar. We recast its well known proof by Picard iteration as the method of power series in order to emphasize the dependence of y_1 and y_2 on q.

For $q = 0$, we have

$$y_1 = \cos \sqrt{\lambda}\, x, \qquad y_2 = \frac{\sin \sqrt{\lambda}\, x}{\sqrt{\lambda}}.$$

These are entire function of λ, because

$$\cos \sqrt{\lambda}\, x = \sum_{n \geq 0} \frac{(-1)^n}{(2n)!} x^{2n} \lambda^n$$

$$\frac{\sin \sqrt{\lambda}\, x}{\sqrt{\lambda}} = \sum_{n \geq 0} \frac{(-1)^n}{(2n + 1)!} x^{2n+1} \lambda^n.$$

This can be generalized.

Analyticity Properties. (a) *For each $x \in [0, 1]$,*

$$y_j(x, \lambda, q), \qquad y_j'(x, \lambda, q), \qquad j = 1, 2$$

are entire functions on $\mathbb{C} \times L_{\mathbb{C}}^2$. They are real valued on $\mathbb{R} \times L_{\mathbb{R}}^2$.

(b) *The solution*

$$y_j(\cdot, \lambda, q), \qquad j = 1, 2$$

is analytic as a map from $\mathbb{C} \times L_{\mathbb{C}}^2$ into $H_{\mathbb{C}}^2$.

For $k \geq 0$ an integer, $H_{\mathbb{C}}^k = H_{\mathbb{C}}^k[0, 1]$ denotes the Hilbert space of all complex valued functions on $[0, 1]$, which have k derivatives in L^2.[6] $H_{\mathbb{R}}^k$ is the subspace of all real valued functions in $H_{\mathbb{C}}^k$. In particular, $L_{\mathbb{C}}^2 = H_{\mathbb{C}}^0$ and $L_{\mathbb{R}}^2 = H_{\mathbb{R}}^0$.

Proof of Analyticity Properties. (a) We consider y_1. The nth term in its power series expansion is

$$C_n = \int_{0 \leq t_1 \leq \cdots \leq t_{n+1} = x} c_\lambda(t_1) \prod_{i=1}^{n} s_\lambda(t_{i+1} - t_i) \prod_{i=1}^{n} q(t_i)\, dt_1 \cdots dt_n.$$

The first two factors in the integrand are entire functions of λ. For each x, this term therefore is continuously differentiable in λ and q everywhere on

[6]For example, f in $L_{\mathbb{C}}^2$ belongs to $H_{\mathbb{C}}^1$, if the difference quotients $(f(x + h) - f(x))/h$ have a strong limit in $L_{\mathbb{C}}^2$.

$\mathbb{C} \times L_{\mathbb{C}}^2$, hence is an entire function of λ and q. By the uniform convergence of the sum of all these terms on bounded subsets of $[0, 1] \times \mathbb{C} \times L_{\mathbb{C}}^2$, y_1 is also an entire function of λ and q for each x.

By the integral equation of Theorem 1,

$$y_1'(x, \lambda, q) = -\sqrt{\lambda} \sin \sqrt{\lambda}\, x + \int_0^x \cos \sqrt{\lambda}(x - t)q(t)y_1(t, \lambda, q)\, dt.$$

This shows that for each x, the derivative y_1' is also continuously differentiable hence entire in λ and q.

By the differential equation,

$$y_1(x, \overline{\lambda}, \overline{q}) = \overline{y_1(x, \lambda, q)},$$

so y_1 and y_1' are real valued for real λ and q.

(b) Again, consider y_1. Each term C_n is in fact continuously differentiable, hence analytic, when considered as a map from $\mathbb{C} \times L_{\mathbb{C}}^2$ into $C_{\mathbb{C}}$, the Banach space of all complex valued continuous functions on $[0, 1]$ with the usual supremum norm. The same is then true for y_1 by uniform convergence, and for y_1' by the integral equation. It follows that y_1 and y_1' are also analytic as maps from $\mathbb{C} \times L_{\mathbb{C}}^2$ into $L_{\mathbb{C}}^2$, since the supremum norm is stronger than the L^2-norm. By the differential equation,

$$y_1'' = (q - \lambda)y_1$$

is analytic in the same sense. This shows that y_1 is analytic as a map from $\mathbb{C} \times L_{\mathbb{C}}^2$ into $H_{\mathbb{C}}^2$. ∎

The *Wronskian* of two differentiable functions f and g is the function

$$[f, g] = \begin{vmatrix} f & g \\ f' & g' \end{vmatrix} = fg' - f'g.$$

We have the very important

Wronskian Identity.

$$[y_1, y_2] = 1.$$

That is,[7]

$$\begin{bmatrix} y_1 & y_2 \\ y_1' & y_2' \end{bmatrix} \in SL(2, \mathbb{C}).$$

---

[7] $SL(2, \mathbb{C})$ is the group of all 2×2 complex matrices with determinant 1.

Proof. We have

$$[y_1, y_2]' = (y_1 y_2' - y_1' y_2)'$$

$$= y_1 y_2'' - y_1'' y_2$$

$$= y_1(q - \lambda)y_2 - (q - \lambda)y_1 y_2$$

$$= 0$$

almost everywhere, hence

$$[y_1, y_2](x) = [y_1, y_2](0) = 1$$

everywhere by continuity. ∎

It is now possible to generalize Lemma 1.

Theorem 2. *Let $f \in L_{\mathbb{C}}^2$ and $a, b \in \mathbb{C}$. Then there exists a unique solution of the inhomogeneous equation*

$$-y'' + q(x)y = \lambda y - f(x), \qquad 0 \le x \le 1,$$

satisfying

$$y(0) = a, \qquad y'(0) = b.$$

It is given by

$$y(x) = ay_1(x) + by_2(x) + \int_0^x (y_1(t)y_2(x) - y_1(x)y_2(t))f(t)\, dt.$$

Proof. The proof is the same as that of Lemma 1. One only has to replace $s_\lambda(x - t) = s_\lambda(x)c_\lambda(t) - s_\lambda(t)c_\lambda(x)$ by $y_1(x)y_2(t) - y_1(x)y_2(t)$ and make use of the Wronskian identity. ∎

Corollary 1. *Every solution of equation (1) is uniquely given by*

$$y(x) = y(0)y_1(x) + y'(0)y_2(x).$$

Moreover, if a solution has a double root in $[0, 1]$, then it vanishes identically.

The initial value problem for equation (1) is now solved.

Proof. The first statement is immediate. The second is easily verified. Suppose

$$\begin{bmatrix} y_1(x) & y_2(x) \\ y_1'(x) & y_2'(x) \end{bmatrix} \begin{bmatrix} y(0) \\ y'(0) \end{bmatrix} = \begin{bmatrix} y(x) \\ y'(x) \end{bmatrix} = 0.$$

Then $\binom{y(0)}{y'(0)} = 0$, because the 2×2-matrix is nonsingular by the Wronskian identity. ∎

Comparing Theorem 2 with Lemma 1, y_1 and y_2 appear as generalizations of the trigonometric functions $\cos \sqrt{\lambda}\, x$ and $\sin \sqrt{\lambda}\, x/\sqrt{\lambda}$. It is indeed often helpful to think of them this way. For large λ, this observation can be made quantitative by improving on the bound in Theorem 1. These asymptotic estimates will be used over and over again.

Theorem 3 (Basic Estimates). *On* $[0, 1] \times \mathbb{C} \times L_{\mathbb{C}}^2$,

$$|y_1(x, \lambda, q) - \cos \sqrt{\lambda}\, x| \leq \frac{1}{|\sqrt{\lambda}|} \exp(|\mathrm{Im}\, \sqrt{\lambda}|x + \|q\| \sqrt{x})$$

$$\left| y_2(x, \lambda, q) - \frac{\sin \sqrt{\lambda}\, x}{\sqrt{\lambda}} \right| \leq \frac{1}{|\lambda|} \exp(|\mathrm{Im}\, \sqrt{\lambda}|x + \|q\| \sqrt{x})$$

and

$$|y_1'(x, \lambda, q) + \sqrt{\lambda} \sin \sqrt{\lambda}\, x| \leq \|q\| \exp(|\mathrm{Im}\, \sqrt{\lambda}|x + \|q\| \sqrt{x})$$

$$|y_2'(x, \lambda, q) - \cos \sqrt{\lambda}\, x| \leq \frac{\|q\|}{|\sqrt{\lambda}|} \exp(|\mathrm{Im}\, \sqrt{\lambda}|x + \|q\| \sqrt{x}).$$

Proof. Clearly,

$$|y_1(x, \lambda, q) - \cos \sqrt{\lambda}\, x| \leq \sum_{n \geq 1} |C_n(x, \lambda, q)|.$$

Each term C_n, $n \geq 1$, contains the factor $s_\lambda(x - t_n)$, which may be bounded by

$$|s_\lambda(t)| = \frac{|\sin \sqrt{\lambda}\, t|}{|\sqrt{\lambda}|} \leq \frac{1}{|\sqrt{\lambda}|} \exp(|\mathrm{Im}\, \sqrt{\lambda}|t), \qquad 0 \leq t \leq 1.$$

Otherwise, we proceed as in the proof of Theorem 1 to obtain the estimate

$$|C_n(x, \lambda, q)| \leq \frac{1}{|\sqrt{\lambda}|} \exp(|\mathrm{Im}\, \sqrt{\lambda}|x) \frac{(\|q\| \sqrt{x})^n}{n!}, \qquad n \geq 1.$$

Summing these majorants and including the term for $n = 0$, we obtain the first inequality. The second one is proven by bounding both the factors $s_\lambda(t_1)$ and $s_\lambda(x - t_n)$ occurring in S_n for $n \geq 1$ as above.

Differentiating the integral equation for y_1 one finds

$$y_1'(x, \lambda, q) = -\sqrt{\lambda} \sin \sqrt{\lambda}\, x + \int_0^x \cos \sqrt{\lambda}\,(x - t)q(t)y_1(t, \lambda, q)\, dt.$$

Using the estimate of Theorem 1 and the inequality $|c_\lambda(t)| \leq \exp(|\mathrm{Im}\,\sqrt{\lambda}|t)$,

$$|y_1'(x, \lambda, q) + \sqrt{\lambda} \sin \sqrt{\lambda}\, x| \leq \exp(|\mathrm{Im}\,\sqrt{\lambda}|x) \int_0^x |q(t)|e^{\|q\|\sqrt{t}}\, dt$$

$$\leq \exp(|\mathrm{Im}\,\sqrt{\lambda}|x + \|q\| \sqrt{x}) \int_0^x |q(t)|\, dt$$

$$\leq \|q\| \exp(|\mathrm{Im}\,\sqrt{\lambda}|x + \|q\| \sqrt{x}),$$

proving the third inequality.

To prove the fourth, show that

$$|y_2(x, \lambda, q)| \leq \frac{1}{|\sqrt{\lambda}|} \exp(|\mathrm{Im}\,\sqrt{\lambda}|x + \|q\| \sqrt{x})$$

and mimic the last paragraph. ∎

Problem 2. The entire function $f(z)$ is of order $\omega < \infty$ if for every $\varepsilon > 0$, but for no $\varepsilon < 0$,

$$\sup_{|z| = r} |f(z)| = O(e^{r^{\omega+\varepsilon}}).$$

It is of type $\tau < \infty$ of order ω if for every $\varepsilon > 0$, but for no $\varepsilon < 0$,

$$\sup_{|z| = r} |f(z)| = O(e^{(\tau+\varepsilon)r^\omega}).$$

Show that the order and type of y_j and y_j', $j = 1, 2$, with respect to λ is $\omega = \frac{1}{2}$, $\tau = x$ respectively.

The proof of Theorem 3 implicitly shows that

$$y_1(x, \lambda) = \cos \sqrt{\lambda}\, x + \sum_{n=1}^N C_n(x, \lambda) + O\left(\frac{\exp(|\mathrm{Im}\,\sqrt{\lambda}|)}{|\sqrt{\lambda}|^{N+1}}\right).$$

As it stands, this estimate is not a very useful one since C_n is rather complicated, even for large $|\lambda|$. A small improvement is possible by splitting terms apart using trigonometric identities. For example, using the identity $2 \sin a \cos b = \sin(a + b) + \sin(a - b)$, we obtain

$$C_1(x, \lambda, q) = \frac{1}{\sqrt{\lambda}} \int_0^x \sin \sqrt{\lambda}(x - t) \cos \sqrt{\lambda} t \, q(t) \, dt$$

$$= \frac{\sin \sqrt{\lambda} x}{2\sqrt{\lambda}} \int_0^x q(t) \, dt + \frac{1}{2\sqrt{\lambda}} \int_0^x \sin \sqrt{\lambda}(x - 2t) q(t) \, dt.$$

But in general, one can't do any better.

The situation is quite different, when q has one or more derivatives. Then we may decompose C_n and refine its pieces by partial integration. If q has one derivative in $L_{\mathbb{C}}^2$, then

$$C_1(x, \lambda, q) = \frac{\sin \sqrt{\lambda} x}{2\sqrt{\lambda}} \int_0^x q(t) \, dt + \frac{\cos \sqrt{\lambda} x}{4\lambda} (q(x) - q(0))$$

$$+ \frac{-1}{4\lambda} \int_0^x \cos \sqrt{\lambda}(x - 2t) q'(t) \, dt,$$

where the last term is $o(\exp(|\mathrm{Im}\, \sqrt{\lambda}|x)/|\lambda|)$ as $|\lambda| \to \infty$.[8] If q has a second derivative in $L_{\mathbb{C}}^2$, we can integrate by parts once more and replace the third term by

$$-\frac{\sin \sqrt{\lambda} x}{8 \sqrt{\lambda}^3} (q'(x) + q'(0)) + \frac{-1}{8 \sqrt{\lambda}^3} \int_0^x \sin \sqrt{\lambda}(x - 2t) q''(t) \, dt.$$

Now, the last term is $o(\exp(|\mathrm{Im}\, \sqrt{\lambda}|x)/|\sqrt{\lambda}|^3)$ as $|\lambda| \to \infty$.[8]

Problem 3. Show that for u in $L_{\mathbb{C}}^2$ and $a > 0$,

$$\int_0^a u(t) \sin \sqrt{\lambda} t \, dt = o(e^{|\mathrm{Im}\, \sqrt{\lambda}|a})\ [9]$$

for $|\lambda| \to \infty$. Note that λ need not tend to infinity along the positive real axis.

[Hint: First, prove the statement for continuously differentiable functions, using partial integration. Then write $u = (u - v) + v$, where v is C^1, and $u - v$ has small L^2-norm.]

---

[8] See the next Problem.

[9] That is,

$$\lim_{|\lambda| \to \infty} e^{-|\mathrm{Im}\, \sqrt{\lambda}|a} \int_0^a u(t) \sin \sqrt{\lambda} t \, dt = 0.$$

In the expansions above, terms are ordered by inverse powers of $\sqrt{\lambda}$, and each term (besides the error term) is a product of expressions in λ and q alone, if we ignore the dependence on x. Thus, it is natural, assuming q to be smooth, to split the first N coefficients C_n up in this way and to collect terms of the same order in $1/\sqrt{\lambda}$. The result of this procedure is a formal asymptotic expansion of y_1 to order N in inverse powers of $\sqrt{\lambda}$ with these special coefficients.

There are more efficient ways to generate this expansion, see for example [Wa]. But in any case, it is a real task to actually calculate the coefficients. Luckily, we don't have to, because the estimates of Theorem 3 meet all of our needs. The next theorem is included to give the expansions for twice differentiable functions.

Theorem 4. *For $q \in H_{\mathbb{C}}^2$,*

$$y_1(x, \lambda, q) = \cos\sqrt{\lambda}\,x + \frac{\sin\sqrt{\lambda}\,x}{2\sqrt{\lambda}} Q(x) + \frac{\cos\sqrt{\lambda}\,x}{4\lambda}(q(x) - q(0) - \tfrac{1}{2}Q^2(x))$$

$$+ O\left(\frac{\exp(|\operatorname{Im}\sqrt{\lambda}|x)}{|\sqrt{\lambda}|^3}\right)$$

and

$$y_2(x, \lambda, q) = \frac{\sin\sqrt{\lambda}\,x}{\sqrt{\lambda}} - \frac{\cos\sqrt{\lambda}\,x}{2\lambda} Q(x) + \frac{\sin\sqrt{\lambda}\,x}{4\sqrt{\lambda}^3}(q(x) + q(0) - \tfrac{1}{2}Q^2(x))$$

$$+ O\left(\frac{\exp(|\operatorname{Im}\sqrt{\lambda}|x)}{|\lambda|^2}\right),$$

where

$$Q(x) = \int_0^x q(t)\, dt.$$

Proof. We have already seen that

$$C_1(x, \lambda, q) = \frac{\sin\sqrt{\lambda}\,x}{2\sqrt{\lambda}} \int_0^x q(t)\, dt + \frac{\cos\sqrt{\lambda}\,x}{4\lambda}(q(x) - q(0))$$

$$- \frac{\sin\sqrt{\lambda}\,x}{8\sqrt{\lambda}^3}(q'(x) + q'(0)) + O\left(\frac{\exp(|\operatorname{Im}\sqrt{\lambda}|x)}{|\sqrt{\lambda}|^3}\right).$$

By the same kind of calculation,

$$C_2(x, \lambda, q) = \frac{1}{\lambda} \int_0^x \sin \sqrt{\lambda}\,(x - t_2)q(t_2) \int_0^{t_2} \sin \sqrt{\lambda}\,(t_2 - t_1)$$

$$\cdot \cos \sqrt{\lambda} t_1\, q(t_1)\, dt_1\, dt_2$$

$$= \frac{1}{2\lambda} \int_0^x \sin \sqrt{\lambda}\,(x - t_2)q(t_2) \sin \sqrt{\lambda}\, t_2 \int_0^{t_2} q(t_1)\, dt_1\, dt_2$$

$$+ \frac{1}{2\lambda} \int_0^x \sin \sqrt{\lambda}\,(x - t_2)q(t_2) \int_0^{t_2} \sin \sqrt{\lambda}\,(t_2 - 2t_1)q(t_1)\, dt_1\, dt_2$$

$$= I + II.$$

Using the identity $2 \sin a \sin b = -\cos(a + b) + \cos(a - b)$ and integration by parts,

$$I = -\frac{\cos \sqrt{\lambda}\, x}{4\lambda} \int_0^x q(t_2) \int_0^{t_2} q(t_1)\, dt_1\, dt_2$$

$$+ \frac{1}{4\lambda} \int_0^x \cos \sqrt{\lambda}\,(x - 2t_2)q(t_2) \int_0^{t_2} q(t_1)\, dt_1\, dt_2$$

$$= -\frac{\cos \sqrt{\lambda}\, x}{4\lambda} \frac{1}{2} \left(\int_0^x q(t)\, dt \right)^2 + O\!\left(\frac{\exp(|\mathrm{Im}\,\sqrt{\lambda}|x)}{|\sqrt{\lambda}|^3} \right).$$

Another partial integration shows that

$$II = O\!\left(\frac{\exp(|\mathrm{Im}\,\sqrt{\lambda}|x)}{|\sqrt{\lambda}|^3} \right).$$

Collecting terms yields the expansion for y_1. The expansion for y_2 follows from similar manipulations. ∎

Problem 4.   Theorem 4 can be improved. For instance, the error term in the expansion of y_2 can be replaced by

$$\frac{\cos \sqrt{\lambda}\, x}{8\lambda^2} \left[q'(x) - q'(0) - (q(x) + q(0))Q(x) - \int_0^x q^2(t)\, dt + \tfrac{1}{6}Q^3(x) \right]$$

$$+ o\!\left(\frac{\exp(|\mathrm{Im}\,\sqrt{\lambda}|x)}{|\lambda|^2} \right).$$

We now investigate the dependence of y_1 and y_2 on q. By construction, they are continuous in the strong topology on $L_{\mathbb{C}}^2$. It turns out, they are even *compact*. That is, continuous with respect to the weak topology.[10]

Theorem 5. *If the sequence q_m converges weakly to q in $L_{\mathbb{C}}^2$, then*

$$y_j(x, \lambda, q_m) \to y_j(x, \lambda, q), \qquad j = 1, 2$$

uniformly on bounded subsets of $[0, 1] \times \mathbb{C}$. In other words, y_1 and y_2 are uniformly compact on bounded subsets of $[0, 1] \times \mathbb{C}$.

This compactness property will ultimately be used to prove existence theorems.

Proof.   Consider y_1. Suppose the sequence q_m converges weakly to q. By the principle of uniform boundedness,

$$\|q\| \le \sup_m \|q_m\| \le M < \infty.$$

If A is any bounded subset of $[0, 1] \times \mathbb{C}$, then

$$|C_n(x, \lambda, p)| \le \frac{1}{n!} \exp(\operatorname{Im} \sqrt{\lambda} \, | \, x)(\|p\| \sqrt{x})^n \le c \frac{M^n}{n!}, \qquad n \ge 1$$

uniformly on A, using the estimate of C_n obtained in the proof of Theorem 1. Thus,

$$
\begin{aligned}
|y_1(x, \lambda, q_m) - y_1(x, \lambda, q)| &\le \sum_{n=1}^{N} |C_n(x, \lambda, q_m) - C_n(x, \lambda, q)| \\
&\quad + \sum_{n=N+1}^{\infty} (|C_n(x, \lambda, q_m)| + |C_n(x, \lambda, q)|) \\
&\le \sum_{n \le 1}^{N} |C_n(x, \lambda, q_m) - C_n(x, \lambda, q)| \\
&\quad + 2c \sum_{n=N+1}^{\infty} \frac{M^n}{n!}
\end{aligned}
$$

uniformly on A. The second sum converges to zero as N tends to infinity. Therefore, it is enough to show that each term $C_n(x, \lambda, q_m)$ converges to $C_n(x, \lambda, q)$ uniformly on A.

---

[10] See Appendix A for compact mappings in general.

Fix $n \geq 1$, and consider the function

$$\Delta_m(x, \lambda) = C_n(x, \lambda, q_m) - C_n(x, \lambda, q).$$

We can write

$$\Delta_m(x, \lambda) = \left\langle p_{x, \lambda}, \prod_{i=1}^{n} \bar{q}_m(t_i) - \prod_{i=1}^{n} \bar{q}(t_i) \right\rangle,$$

where

$$p_{x, \lambda}(t_1, \ldots, t_n) = c_\lambda(t_1) \prod_{i=1}^{n} s_\lambda(t_{i+1} - t_i) \mathbb{1}_{\{0 \leq t_1 \leq \cdots \leq t_{n+1} = x\}},$$

and $\mathbb{1}_{\{\cdots\}}$ is the indicator function of the set $\{\cdots\}$. The inner product is taken in the space $L_{\mathbb{C}}^2([0, 1]^n)$.

We have to show that

$$\sup_A |\Delta_m(x, \lambda)| \to 0 \qquad \text{as} \quad m \to \infty.$$

By continuity, the supremum of $|\Delta_m|$ is attained at some point (x_m, λ_m) in the compact closure of A for each m. So equivalently we have to show that

$$|\Delta_m(x_m, \lambda_m)| \to 0 \qquad \text{as} \quad m \to \infty.$$

Suppose this is not the case. Passing to a subsequence we can assume that (x_m, λ_m) converges to (x_*, λ_*), while

$$|\Delta_m(x_m, \lambda_m)| \geq \delta > 0.$$

By the bounded convergence theorem,

$$p_{x_m, \lambda_m} \xrightarrow[\text{strong}]{} p_{x_*, \lambda_*}$$

in $L_{\mathbb{C}}^2([0, 1]^n)$. On the other hand,

$$\prod_{i=1}^{n} q_m(t_i) \xrightarrow[\text{weak}]{} \prod_{i=1}^{n} q(t_i)$$

in $L_{\mathbb{C}}^2([0, 1]^n)$. This is easily checked by integrating against the monomials $t_1^{k_1} \cdots t_n^{k_n}$, which span a dense subspace in $L_{\mathbb{C}}^2([0, 1]^n)$. It follows that[11]

$$|\Delta_m(x_m, \lambda_m)| \to 0,$$

which is a contradiction.

[11] In general, if $f_m \to f$ strongly and $g_m \to g$ weakly in a Hilbert space, then

$$\langle f_m, g_m \rangle = \langle f_m - f, g_m \rangle + \langle f, g_m \rangle \to \langle f, g \rangle,$$

since $\langle f_m - f, g_m \rangle \to 0$ by the Schwarz inequality and the uniform boundedness of the g_m.

This proves the theorem for y_1. The proof for y_2 is of course the same. ∎

The functions y_1 and y_2 are analytic in q. In particular, they are continuously differentiable in q. So, what are their derivatives and gradients?

Recall that the derivative of a map $f: E \rightarrow F$ between two Banach spaces E and F at a point x is a bounded linear map from E into F, which we denote by $d_x f$.[12] Moreover, if E is a Hilbert space and F the real or complex line, then, by the Riesz representation theorem, there is a unique element $\partial f / \partial x$ in E, such that

$$d_x f(v) = \left\langle v, \overline{\frac{\partial f}{\partial x}} \right\rangle$$

for all v in E. This element is the gradient of f at x.

To get an idea what the gradients of y_1 and y_2 look like, we make a formal calculation. Differentiating both sides of the equation

$$-y_j'' + q(x)y_j = \lambda y_j$$

with respect to q in the direction v in $L_\mathbb{C}^2$, we obtain

$$-d_q y_j''(v) + q(x)\, d_q y_j(v) = \lambda d_q y_j(v) - v y_j.$$

Formally, interchanging differentiation with respect to x and q, the first term becomes $-(d_q y_j(v))''$. The initial values of $d_q y_j(v)$ both vanish, since the initial values of y_j are independent of q. Thus, Theorem 2 yields

$$d_q y_j(v) = \int_0^x y_j(t)[y_1(t)y_2(x) - y_1(x)y_2(t)]v(t)\, dt,$$

and consequently

$$\frac{\partial y_j}{\partial q(t)} = y_j(t)[y_1(t)y_2(x) - y_1(x)y_2(t)]\, \mathbb{1}_{[0,\, x]}(t),$$

where $\mathbb{1}_{[0,\, x]}$ is the indicator function of the interval $[0, x]$.

This is easy to make rigorous.

---

[12] See Appendix A for the definition of derivative.

Theorem 6. *For $j = 1, 2$,*

(a)
$$\frac{\partial y_j}{\partial q(t)}(x) = y_j(t)[y_1(t)y_2(x) - y_1(x)y_2(t)] \, \mathbb{1}_{[0,x]}(t)$$

$$\frac{\partial y_j'}{\partial q(t)}(x) = y_j(t)[y_1(t)y_2'(x) - y_1'(x)y_2(t)] \, \mathbb{1}_{[0,x]}(t).$$

The gradients are jointly continuous with respect to x, λ, q.

(b)
$$\frac{\partial y_j}{\partial \lambda} = -\int_0^1 \frac{\partial y_j}{\partial q(t)} \, dt,$$

$$\frac{\partial y_j'}{\partial \lambda} = -\int_0^1 \frac{\partial y_j'}{\partial q(t)} \, dt.$$

Proof. (a) If q is continuous, then y_j is twice continuously differentiable in x, and interchanging differentiation with respect to x and q is allowed. That is,

$$d_q y''_j(v) = (d_q y_j(v))''.$$

Therefore, the result

$$d_q y_j(v) = \int_0^x y_j(t)[y_1(t)y_2(x) - y_1(x)y_2(t)]v(t) \, dt$$

of the preceding calulations indeed holds for continuous functions q.

Now, each side of this equation is a continuous function of q in L_C^2, and the equation holds on a dense subset of L_C^2, namely the continuous functions. Hence it must hold in general. The formula for the gradient follows. Differentiating both sides with respect to x,

$$d_q y_j'(v) = \int_0^x y_j(t)[y_1(t)y_2'(x) - y_1'(x)y_2(t)]v(t) \, dt$$

for continuous q. Again, this holds in general by continuity in q.
   (b) follows from the identity

$$y_j(x, \lambda + \varepsilon, q) = y_j(x, \lambda, q - \varepsilon),$$

which implies

$$\frac{\partial}{\partial \lambda} y_j(x, \lambda, q) = \frac{\partial}{\partial \varepsilon} y_j(x, \lambda, q - \varepsilon)\Big|_{\varepsilon = 0}$$

$$= -\int_0^1 \frac{\partial y_j}{\partial q(t)} \, dt.$$

Thus, $\partial y_j/\partial\lambda$ is just the directional derivative of y_j at q in the direction of the constant function -1. Same for $\partial y_j'/\partial\lambda$.   ■

Part (a) of Theorem 6 can be written more compactly as

$$\frac{\partial}{\partial q(t)}\begin{bmatrix} y_1(x) & y_2(x) \\ y_1'(x) & y_2'(x) \end{bmatrix}$$

$$= \begin{bmatrix} y_1(x) & y_2(x) \\ y_1'(x) & y_2'(x) \end{bmatrix}\begin{bmatrix} -y_1(t)y_2(t) & -y_2^2(t) \\ y_1^2(t) & y_1(t)y_2(t) \end{bmatrix}\mathbb{1}_{[0,x]}(t).$$

Problem 5.   To interpret the last identity, fix x and λ, and consider the map

$$q \rightarrow Y(x, \lambda, q) = \begin{bmatrix} y_1 & y_2 \\ y_1' & y_2' \end{bmatrix}$$

from $L_{\mathbb{C}}^2$ into $SL(2, \mathbb{C})$. The tangent space to $SL(2, \mathbb{C})$ at Y_0 is the three complex dimensional space of matrices of the form $Y_0 X$, where X is in $sl(2, \mathbb{C})$, the Lie algebra of all complex 2×2-matrices with trace zero. Show that the tangent vector to the curve $Y(x, \lambda, q + tv)$ at $t = 0$ is given by the above identity.

Problem 6.   Prove directly that

$$\|y_j(x, \lambda, q + v) - y_j(x, \lambda, q) - L_q(v)\| = o(\|v\|),$$

where

$$L_q(v) = \int_0^x y_j(t)[y_1(t)y_2(x) - y_1(x)y_2(t)]v \, dt.$$

[Hint: Observe that $y_j(x, \lambda, q + v)$ satisfies an inhomogeneous equation with inhomogeneity $vy_j(x, \lambda, q + v)$, and apply Theorem 2.]

Problem 7.   Are there critical points of the functions y_1 and y_2 on $L_{\mathbb{C}}^2$? That is, regard x and λ as fixed, and determine the set of qs for which the gradient $\partial y_j/\partial q(t)$ vanishes identically in t.

One interesting feature of Theorem 6 is that the derivative of y_1 and y_2 with respect to q is expressed in terms of products of y_1 and y_2. This phenomenon will occur many times. We end this chapter with a remark about such products.

Theorem 7.   (a)   *For each* $(\lambda, q) \in \mathbb{C} \times L^2_{\mathbb{C}}$, *the functions*

$$y_1^2, y_1 y_2, y_2^2$$

are linearly independent over $[0, 1]$.
(b)   *Let* $q \in C^1_{\mathbb{C}}[0, 1]$, *and*

$$L = q(x)\frac{d}{dx} + \frac{d}{dx}q(x) - \frac{1}{2}\left(\frac{d}{dx}\right)^3.$$

If f, g *are two solutions of equation* (1) *for the same* λ, *then*

$$L(fg) = 2\lambda \frac{d}{dx}(fg).$$

Proof.   (a)   Substracting $2(q - \lambda)$ times the first row from the third row and using the differential equation, we get

$$\begin{vmatrix} y_1^2 & y_1 y_2 & y_2^2 \\ (y_1^2)' & (y_1 y_2)' & (y_2^2)' \\ (y_1^2)'' & (y_1 y_2)'' & (y_2^2)'' \end{vmatrix} = \begin{vmatrix} y_1^2 & y_1 y_2 & y_2^2 \\ 2y_1 y_1' & y_1 y_2' + y_1' y_2 & 2y_2 y_2' \\ 2(y_1')^2 & 2y_1' y_2' & 2(y_2')^2 \end{vmatrix}$$

$$= 2(y_1 y_2' - y_1' y_2)^3$$

$$= 2$$

almost everywhere by the Wronskian identity. Hence, $y_1^2, y_1 y_2, y_2^2$ are linearly independent on $[0, 1]$.
(b)   follows by direction calculation.   ■

Problem 8.   (a)   Check that the map

$$g = \begin{pmatrix} a & b \\ c & d \end{pmatrix} \rightarrow$$

$$R_g = \begin{pmatrix} \frac{1}{2}(a^2 + b^2 + c^2 + d^2) & \frac{1}{2}(a^2 - b^2 + c^2 - d^2) & ab + cd \\ \frac{1}{2}(a^2 + b^2 - c^2 - d^2) & \frac{1}{2}(a^2 - b^2 - c^2 + d^2) & ab - cd \\ ac + bd & ac - bd & ad + bc \end{pmatrix}$$

is a homomorphism of $SL(2, \mathbb{C})$ into $SO(1, 2, \mathbb{C})$, the group of all 3×3-matrices with complex elements and determinant 1, which preserve the quadratic form $x^2 - y^2 - z^2$.

(b) Let $\phi = (\phi_1, \phi_2)$ be any fundamental solution of equation (1) with $[\phi_1, \phi_2] = 1$. Set

$$\Phi_\phi = [\tfrac{1}{2}(\phi_1^2 + \phi_2^2), \tfrac{1}{2}(\phi_1^2 - \phi_2^2), \phi_1\phi_2].$$

Verify the identity

$$R_g\Phi_\phi = \Phi_{g\phi}.$$

(c) Prove that there exists a $g \in SL(2, \mathbb{C})$ such that

$$\langle \Phi_1, \Phi_j \rangle = \delta_{ij}, \qquad 1 \le i, j \le 3,$$

where $(\Phi_1, \Phi_2, \Phi_3) = \Phi_{g\phi}$.

2 The Dirichlet Problem

This chapter is devoted to the simplest boundary value problem for the differential equation

(1) $$-y'' + q(x)y = \lambda y, \qquad 0 \leq x \leq 1.$$

Here, $\lambda \in \mathbb{C}$ and $q \in L^2 = L^2_{\mathbb{R}}[0, 1]$, the Hilbert space of all *real* valued, square integrable functions on $[0, 1]$. We ask whether there are nontrivial solutions of equation (1) satisfying the Dirichlet boundary condition

$$y(0) = 0, \qquad y(1) = 0.$$

This is the Dirichlet problem.

A complex number λ is called a Dirichlet eigenvalue of q if the Dirichlet problem can be solved. The corresponding nontrivial solutions are called eigenfunctions of q for λ. The collection of all eigenvalues of q is its Dirichlet spectrum.

From a more abstract point of view, the Dirichlet problem belongs to the theory of self-adjoint operators. Precisely, the unbounded operator Q, defined by

$$Qf = \left(-\frac{d^2}{dx^2} + q \right) f$$

for f in the dense subspace D of all functions in $H^2_{\mathbb{C}}[0, 1]$ that vanish at 0 and 1, is self-adjoint on the domain D.[1] It can be shown that the spectrum of Q coincides with the Dirichlet spectrum of q, and so we may draw some conclusions about the Dirichlet problem from general facts about self-adjoint operators. For instance, the Dirichlet spectrum must be an unbounded sequence of real numbers. However, abstract operator theory is not helpful in answering many of the questions that interest us. For this reason, we shall follow Sturm's classical approach [St] which may be called the "method of shooting."

Intuitively speaking, the solution y_2 is shot from the left end point of the interval with unit velocity. At $x = 1$, it reaches the height $y_2(1, \lambda)$. The idea is to vary the parameter λ so that the terminal height is 0. This point is illustrated in Figure 1. At the moment, the terminal velocity $y'_2(1, \lambda)$ does not concern us. It becomes important in Chapter 3.

The "shooting method" is easy to implement. If μ is a zero of $y_2(1, \lambda)$, then $y_2(x, \mu)$ is a nontrivial solution of equation (1) satisfying the Dirichlet boundary conditions. Hence, μ is a Dirichlet eigenvalue of q. Conversely, suppose μ is a Dirichlet eigenvalue of q with eigenfunction $y(x)$. Then

$$y(x) = y'(0)y_2(x, \mu)$$

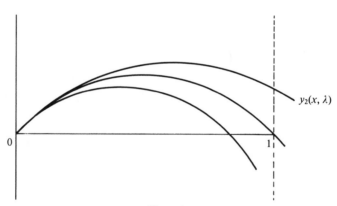

$$y_2(x, \lambda)$$

Figure 1.

[1] That is, for all $f, g \in D$,

$$\langle Qf, g \rangle = \langle f, Qg \rangle,$$

and for all $g \in L^2_{\mathbb{C}}$, the condition

$$\sup_{f \in D, \|f\| = 1} |\langle Qf, g \rangle| < \infty$$

implies that g is in D. See [RS].

by Corollary 1.1, hence $y_2(1, \mu) = 0$ by the boundary conditions. We see that the Dirichlet spectrum of q is identical with the zero set of the entire function $y_2(1, \lambda, q)$. From now on, we will therefore not distinguish between them.

The spectrum of a finite dimensional linear map is the zero set of its characteristic polynomial. By analogy, we can consider $y_2(1, \lambda, q)$ as the "characteristic polynomial" of the linear operator $-d^2/dx^2 + q$ with Dirichlet boundary conditions.

For $q = 0$, the Dirichlet spectrum is the infinite sequence

$$\pi^2, 4\pi^2, 9\pi^2, \ldots, n^2\pi^2, \ldots,$$

since $y_2(1, \lambda, 0) = \sin \sqrt{\lambda}/\sqrt{\lambda}$. The "Counting Lemma" below shows that there are infinitely many eigenvalues for every q. It also gives a first rough estimate of their location, to be refined later on.

First, a simple technical lemma.

Lemma 1. *If $|z - n\pi| \geq \pi/4$ for all integers n, then*

$$e^{|\operatorname{Im} z|} < 4|\sin z|.$$

Proof. Write $z = x + iy$ with real x, y. Since $|\sin z|$ is even and periodic with period π, it suffices to prove the lemma for $0 \leq x \leq \pi/2$ and $|z| \geq \pi/4$.

We have

$$|\sin z|^2 = \cosh^2 y - \cos^2 x.$$

For $\pi/6 \leq x \leq \pi/2$,

$$\cos^2 x \leq \tfrac{3}{4} \leq \tfrac{3}{4} \cosh^2 y$$

for all real y. For $0 \leq x \leq \pi/6$, the assumption $|z| \geq \pi/4$ implies $y^2 \geq (\pi/4)^2 - x^2 \geq \tfrac{5}{144}\pi^2 \geq \tfrac{1}{3}$ and hence

$$\cosh^2 y \geq 1 + y^2 \geq \tfrac{4}{3} \geq \tfrac{4}{3} \cos^2 x$$

as before. Thus, in both cases we have

$$|\sin z|^2 \geq \tfrac{1}{4} \cosh^2 y > \tfrac{1}{16} e^{2|y|},$$

and the result follows. ∎

The estimate of Lemma 1 is not optimal, but chosen for the sake of convenience. In general, there is a positive constant c_δ for each $\delta > 0$ such that

$$e^{|\operatorname{Im} z|} < c_\delta |\sin z|$$

provided $|z - n\pi| > \delta$ for all n.

In the next lemma, we admit complex valued functions q. This is useful, for instance in the proof of Theorem 3.1.

Lemma 2 (The Counting Lemma). *Let $q \in L_{\mathbb{C}}^2$ and $N > 2e^{\|q\|}$ be an integer. Then $y_2(1, \lambda, q)$ has exactly N roots, counted with multiplicities, in the open half plane*

$$\operatorname{Re} \lambda < (N + \tfrac{1}{2})^2 \pi^2,$$

and for each $n > N$, exactly one simple root in the egg shaped region

$$|\sqrt{\lambda} - n\pi| < \frac{\pi}{2}.$$

There are no other roots.

Proof. Fix $N > 2e^{\|q\|}$, and let $K > N$ be another integer. Consider the contours

$$|\sqrt{\lambda}| = (K + \tfrac{1}{2})\pi,$$

$$\operatorname{Re} \sqrt{\lambda} = (N + \tfrac{1}{2})\pi,$$

and

$$|\sqrt{\lambda} - n\pi| = \frac{\pi}{2}, \qquad n > N.$$

See Figure 2. By Lemma 1, the estimate

$$e^{|\operatorname{Im} \sqrt{\lambda}|} < 4|\sin \sqrt{\lambda}|$$

holds on all of them. Therefore, by the Basic Estimate for y_2,

$$\left| y_2(1, \lambda) - \frac{\sin \sqrt{\lambda}}{\sqrt{\lambda}} \right| \le \frac{e^{\|q\|}}{|\sqrt{\lambda}|} \frac{e^{|\operatorname{Im} \sqrt{\lambda}|}}{|\sqrt{\lambda}|}$$

$$< \frac{2N}{|\sqrt{\lambda}|} \left| \frac{\sin \sqrt{\lambda}}{\sqrt{\lambda}} \right|$$

$$< \left| \frac{\sin \sqrt{\lambda}}{\sqrt{\lambda}} \right|$$

also holds on them. It follows that $y_2(1, \lambda)$ does not vanish on these contours. Hence, by Rouché's theorem, $y_2(1, \lambda)$ has as many roots, counted with multiplicities, as $\sin \sqrt{\lambda}/\sqrt{\lambda}$ in each of the bounded regions and the remaining

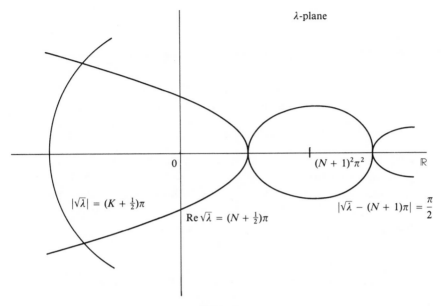

Figure 2.

unbounded region. Since $\sin\sqrt{\lambda}/\sqrt{\lambda}$ has only the simple roots $n^2\pi^2$, $n \geq 1$, and since $K > N$ can be chosen arbitrarily large, the lemma follows. ∎

Returning to real valued functions q, we have

Theorem 1. *The Dirichlet spectrum of q in $L^2_{\mathbb{R}}$ is an infinite sequence of real numbers, which is bounded from below and tends to $+\infty$.*

Proof. We only have to prove reality, the rest follows from the Counting Lemma.

Suppose λ is a Dirichlet eigenvalue of q with eigenfunction $y(x)$. Then

$$-y'' + q(x)y = \lambda y.$$

Conjugating the equation,

$$-\bar{y}'' + q(x)\bar{y} = \bar{\lambda}\bar{y},$$

since q is *real*. Multiplying the first equation by $\bar{y}$, the second by y and taking the difference, we obtain

$$[y, \bar{y}]' = y\bar{y}'' - y''\bar{y} = (\lambda - \bar{\lambda})|y|^2,$$

hence, by integration,

$$\left. [y, \bar{y}] \right|_0^1 = (\lambda - \bar{\lambda}) \int_0^1 |y|^2 \, dt.$$

The left hand side vanishes, while the integral does not. Therefore $\lambda - \bar{\lambda} = 0$, that is, λ is real. Of course, this is the standard argument, used to prove that eigenvalues of self-adjoint operators are real. ∎

It has been shown that every eigenfunction for an eigenvalue λ is a multiple of $y_2(x, \lambda)$. The geometric multiplicity of λ, which is the maximal number of linearly independent eigenfunctions for λ, is therefore 1. On the other hand, its algebraic multiplicity, defined as its order as a root of $y_2(1, \lambda)$, might very well be larger. This, however, is not the case.

We will often use the abbreviated notation $\cdot = \partial/\partial\lambda$.

Theorem 2. *If λ is a Dirichlet eigenvalue of q in L^2, then*

$$\dot{y}_2(1, \lambda) y_2'(1, \lambda) = \int_0^1 y_2^2(t, \lambda) \, dt$$

$$= \| y_2(\cdot, \lambda) \|^2 > 0.$$

In particular, $\dot{y}_2(1, \lambda) \neq 0$. Thus, all roots of $y_2(1, \lambda)$ are simple.

Proof. Let $y_2 = y_2(x, \lambda)$. Differentiating equation (1) with respect to λ yields[2]

$$-\ddot{y}_2'' + q(x)\dot{y}_2 = y_2 + \lambda\dot{y}_2.$$

Multiplying this equation by y_2, the original equation by $\dot{y}_2$ and taking the difference, we obtain

$$y_2^2 = y_2'' \dot{y}_2 - \ddot{y}_2'' y_2 = [\dot{y}_2, y_2]'.$$

Hence

$$\int_0^1 y_2^2(t, \lambda) \, dt = \left. [\dot{y}_2, y_2] \right|_0^1$$

$$= \dot{y}_2(1, \lambda) y_2'(1, \lambda),$$

---

[2] To avoid the problem of interchanging x- and λ-derivative, first assume q is continuous and argue as in the proof of Theorem 1.6.

since $y_2(0, \lambda)$ and $\dot{y}_2(0, \lambda)$ vanish for all λ, and $y_2(1, \lambda)$ vanishes for a Dirichlet eigenvalue λ. The integral is equal to $\| y_2(\cdot, \lambda) \|^2$, since y_2 is real for real λ. ∎

Problem 1. Prove the identity of Theorem 2 by starting from

$$\dot{y}_2(1, \lambda) = -\int_0^1 \frac{\partial y_2(1)}{\partial q(t)} dt$$

and using Theorem 1.6.

It follows from the preceeding results that the Dirichlet spectrum of every q in L^2 is a sequence of real numbers

$$\mu_1(q) < \mu_2(q) < \cdots$$

satisfying

$$\mu_n(q) = n^2\pi^2 + O(n).$$

To each eigenvalue we associate a unique eigenfunction $g_n = g_n(x, q)$ normalized by

$$\|g_n\| = 1, \qquad g_n'(0) > 0.$$

By Theorem 2,

$$g_n(x, q) = \frac{y_2(x, \mu_n)}{\| y_2(\cdot, \mu_n) \|}$$

$$= \frac{y_2(x, \mu_n)}{\sqrt{\dot{y}_2(1, \mu_n) y_2'(1, \mu_n)}},$$

where the argument q has been suppressed on the right hand side. In particular,

$$g_n(x, 0) = \sqrt{2} \sin \pi n x.$$

Let us investigate the Dirichlet eigenvalues as functions defined on L^2.

Theorem 3. μ_n, $n \geq 1$, *is a compact, real analytic function on L^2. Its gradient is*

$$\frac{\partial \mu_n}{\partial q(t)} = g_n^2(t, q).$$

Since μ_n is real analytic, g_n is also a real analytic function of q by the expression above and the Analyticity Properties of Chapter 1.

Proof. To verify compactness, suppose the sequence q_m, $m \geq 1$, converges weakly to q. By the principle of uniform boundedness,

$$\|q\| \leq \sup_m \|q_m\| \leq M < \infty.$$

Let $N > 2e^M$, $\varepsilon > 0$, and consider the intervals

$$I_n = \{\lambda \in \mathbb{R} : |\lambda - \mu_n(q)| < \varepsilon\}, \qquad 1 \leq n \leq N.$$

If ε is sufficiently small, then these intervals are all disjoint and contained in the half line $(-\infty, (N + \frac{1}{2})^2\pi^2)$ by the Counting Lemma. Moreover, $y_2(1, \lambda, q)$ changes sign on each of them, since $\mu_n(q)$ is a simple root.

As m tends to infinity, the functions $y_2(1, \lambda, q_m)$ converge to $y_2(1, \lambda, q)$ uniformly on $I_1 \cup \cdots \cup I_N$ by Theorem 1.5. Hence, for sufficiently large m, they also change sign on $I_1, \ldots, I_N$, so they must all have at least one root in each of these intervals. But there are only N roots on the whole half line $(-\infty, (N + \frac{1}{2})^2\pi^2)$ by the Counting Lemma. Therefore, $y_2(1, \lambda, q_m)$ has exactly one root in each interval I_n, $1 \leq n \leq N$, which must be the nth eigenvalue of q_m. Consequently,

$$|\mu_n(q_m) - \mu_n(q)| < \varepsilon, \qquad 1 \leq n \leq N,$$

for all sufficiently large m. It follows that

$$\mu_n(q_m) \to \mu_n(q)$$

for $m \to \infty$, since N and $\varepsilon > 0$ were arbitrary. Thus, μ_n is a compact function of q.

To prove real analyticity, fix p in L^2. By Theorem 2,

$$\dot{y}_2(1, \mu_n(p), p) \neq 0.$$

So the implicit function theorem applies, and there exists a *unique* continuous function $\hat{\mu}_n$, defined on some small neighborhood $U \subset L^2$ of p, such that

$$y_2(1, \hat{\mu}_n(q), q) = 0, \qquad \hat{\mu}_n(p) = \mu_n(p)$$

on U. Furthermore, $\hat{\mu}_n$ is real analytic. On the other hand, μ_n is also a continuous function on U satisfying $y_2(1, \mu_n(q), q) = 0$. Therefore,

$$\hat{\mu}_n(q) = \mu_n(q)$$

on U by uniqueness, and so μ_n is real analytic.

To calculate the gradient observe that

$$0 = \frac{\partial}{\partial q} y_2(1, \mu_n(q), q)$$

$$= \dot{y}_2(1, \mu_n) \frac{\partial \mu_n}{\partial q} + \frac{\partial}{\partial q} y_2(1, \lambda) \Big|_{\lambda = \mu_n}.$$

By Theorem 1.6, the second term equals

$$y_2(t)[y_1(t)y_2(1) - y_1(1)y_2(t)] = -y_1(1)y_2^2(t)$$

$$= -\frac{1}{y_2'(1)} y_2^2(t),$$

since $y_2(1, \mu_n) = 0$ and $y_1(1)y_2'(1) = 1$ by the Wronskian identity. Hence,

$$\frac{\partial \mu_n}{\partial q(t)} = \frac{y_2^2(t, \mu_n)}{\dot{y}_2(1, \mu_n)y_2'(1, \mu_n)} = g_n^2(t, q). \quad \blacksquare$$

Here is another, perhaps more intuitive proof of the last identity. Differentiating both sides of the differential equation of g_n in the direction v we obtain

$$-d_q g_n''(v) + q d_q g_n(v) + v g_n = \mu_n d_q g_n(v) + d_q \mu_n(v) g_n.$$

If q is continuous, then g_n is twice continuously differentiable, and we may interchange differentiation with respect to x and q to obtain

$$-(d_q g_n(v))'' + q d_q g_n(v) + v g_n = \mu_n d_q g_n(v) + d_q \mu_n(v) g_n.$$

Multiplying both sides by g_n and integrating we find

$$\langle Q d_q g_n(v), g_n \rangle + \langle g_n^2, v \rangle = \mu_n \langle d_q g_n(v), g_n \rangle + d_q \mu_n(v)$$

where $Q = -d^2/dx^2 + q$. The first term equals

$$\langle d_q g_n(v), Q g_n \rangle = \mu_n \langle d_q g_n(v), g_n \rangle.$$

Hence

$$d_q \mu_n(v) = \langle g_n^2, v \rangle,$$

and

$$\frac{\partial \mu_n}{\partial q(t)} = g_n^2(t, q).$$

However, both sides are continuous functions of q, and the continuous q are dense in L^2. So this identity holds in general.

Problem 2. Show that

$$|\mu_n(q) - \mu_n(p)| \le \|q - p\|_\infty, \qquad n \ge 1,$$

if q, p are in L^∞, the space of all real valued, essentially bounded functions on $[0, 1]$.

[Hint: Write

$$\mu_n(q) - \mu_n(p) = \int_0^1 \frac{d}{dt} \mu_n(tq + (1 - t)p) \, dt.]$$

We make a small digression to illustrate the usefulness of Theorem 3. Consider the restriction of the function μ_n, $n \ge 1$, to the sphere $\|q\| = 1$ in L^2. The ball $\|q\| \le 1$ is compact in the weak topology of L^2,[3] and μ_n is a compact function on L^2. Hence, μ_n attains its maximum and minimum on $\|q\| \le 1$. However, we have

$$\mu_n(q + c) = \mu_n(q) + c$$

for real c by the differential equation. It follows that the maximum and minimum must be attained on the unit sphere $\|q\| = 1$, for otherwise they would not be extremal. Consequently, μ_n has at least two critical points on the unit sphere in L^2.

Let q be a critical point of μ_n on $\|q\| = 1$. The gradient of μ_n must be proportional to the unit normal at q. That is,

(*) $$\frac{\partial \mu_n}{\partial q(t)} = g_n^2(t, q) = \alpha q(t),$$

where

$$\alpha = \left(\int_0^1 q(t) \, dt \right)^{-1} \ne 0.$$

By (*), q has exactly $n + 1$ double roots in $[0, 1]$ including 0 and 1, and does not change sign. Moreover, q is infinitely often differentiable. For, g_n^2 is always continuous, hence so is q. But then g_n^2 is in C^1, hence so is q, ad infinitum.

The critical point q also satisfies a differential equation. Apply the operator L of Theorem 1.7 to (*) to obtain

$$-2\mu_n q' + 3qq' - \tfrac{1}{2}q''' = 0.$$

---

[3] Recall that the sphere $\|q\| = 1$ is not compact.

Integration, multiplication by q' and another integration yields the identity

$$-\mu_n q^2 + \tfrac{1}{2} q^3 - \tfrac{1}{4}(q')^2 = cq$$

with some constant $c \neq 0$. The substitution $q = 2u + \tfrac{2}{3}\mu_n$ then leads to the differential equation

(**)
$$(u')^2 = 4(u - e_1)(u - e_2)(u - e_3)$$

where e_1, e_2, e_3 are complex numbers whose sum is zero. For distinct e_1, e_2, e_3, (**) is the differential equation for the Weierstrass $\wp$-function.

It is possible to carry this analysis further using properties of the Weierstrass $\wp$-function. But we shall not.

The Counting Lemma gave us a first rough estimate of the location of the Dirichlet eigenvalues. This estimate is now refined.

Let ℓ^2 denote the Hilbert space of all real sequences $(\alpha_1, \alpha_2, \ldots)$ such that $\sum \alpha_n^2$ is finite. More generally, for $k \geq 0$, ℓ_k^2 denotes the Hilbert space of all real sequences $(\alpha_1, \alpha_2, \ldots)$ such that

$$\sum_{n \geq 1} (n^k \alpha_n)^2 < \infty.$$

In analogy to the notation $O(1/n)$, we use the notation $\ell_k^2(n)$ for an arbitrary sequence of numbers which is an element of ℓ_k^2. For instance,

$$\alpha_n = \beta_n + \ell^2(n)$$

is equivalent to

$$\alpha_n = \beta_n + \lambda_n, \qquad \sum_{n \geq 1} \lambda_n^2 < \infty.$$

Theorem 4. *For q in L^2,*

$$\mu_n(q) = n^2\pi^2 + \int_0^1 q(t)\, dt - \langle \cos 2\pi nx, q \rangle + O\!\left(\frac{1}{n}\right)$$

$$= n^2\pi^2 + \int_0^1 q(t)\, dt + \ell^2(n)$$

and

$$g_n(x, q) = \sqrt{2} \sin \pi nx + O\!\left(\frac{1}{n}\right)$$

$$g_n'(x, q) = \sqrt{2}\, \pi n \cos \pi nx + O(1).$$

These estimates hold uniformly on bounded subsets of $[0, 1] \times L^2$.

Proof. We "iterate" estimates on μ_n and g_n, thereby sharpening them. All estimates hold uniformly on bounded subsets of $[0, 1] \times L^2$.

Let $\mu_n = \mu_n(q)$. By the Counting Lemma,

$$\sqrt{\mu_n} = n\pi + O(1).$$

By the Basic Estimate for y_2,

$$y_2(x, \mu_n) = \frac{\sin \sqrt{\mu_n}\,x}{\sqrt{\mu_n}} + O\left(\frac{e^{\|q\|}}{|\mu_n|}\right)$$

$$= \frac{\sin \sqrt{\mu_n}\,x}{\sqrt{\mu_n}} + O\left(\frac{1}{n^2}\right).$$

Using the identity $2 \sin^2 ax = 1 - \cos 2ax$, one calculates

$$\int_0^1 y_2^2(t, \mu_n)\,dt = \int_0^1 \frac{\sin^2\sqrt{\mu_n}\,t}{\mu_n}\,dt + O\left(\frac{1}{n^3}\right)$$

$$= \frac{1}{2\mu_n}\left(1 - \frac{\sin 2\sqrt{\mu_n}}{2\sqrt{\mu_n}}\right) + O\left(\frac{1}{n^3}\right)$$

$$= \frac{1}{2\mu_n}\left(1 + O\left(\frac{1}{n}\right)\right).$$

It follows that

$$\|y_2(\cdot, \mu_n)\|^{-1} = \sqrt{2} \cdot \sqrt{\mu_n}\left(1 + O\left(\frac{1}{n}\right)\right).$$

Thus the preliminary estimate

(2) $$g_n(x) = \frac{y_2(x, \mu_n)}{\|y_2(\cdot, \mu_n)\|} = \sqrt{2} \sin \sqrt{\mu_n}\,x + O\left(\frac{1}{n}\right).$$

is obtained.

Our estimate of μ_n may now be improved. By Theorem 3,

$$\mu_n - n^2\pi^2 = \mu_n(q) - \mu_n(0)$$

(3) $$= \int_0^1 \frac{d}{dt}\mu_n(tq)\,dt$$

$$= \int_0^1 \langle g_n^2(x, tq), q \rangle\,dt.$$

Since (2) holds uniformly for tq, $0 \le t \le 1$, this implies

$$\mu_n = n^2\pi^2 + O(1)$$

or equivalently

$$\sqrt{\mu_n} = n\pi + O\left(\frac{1}{n}\right),$$

and in turn

$$g_n(x) = \sqrt{2}\sin \pi nx + O\left(\frac{1}{n}\right).$$

Inserting the last estimate of g_n into (3) and using the identity $2\sin^2 ax = 1 - \cos 2ax$ again, we obtain

$$\mu_n - n^2\pi^2 = \int_0^1 \left\langle 1 - \cos 2\pi nx + O\left(\frac{1}{n}\right), q \right\rangle dt$$

$$= \int_0^1 q(t)\, dt - \langle \cos 2\pi nx, q \rangle + O\left(\frac{1}{n}\right)$$

$$= \int_0^1 q(t)\, dt + \ell^2(n),$$

since $\langle \cos 2\pi nx, q \rangle$ are the square summable Fourier cosine coefficients of q.

Finally, we estimate g_n'. By the Basic Estimates,

$$y_2'(x, \mu_n) = \cos \sqrt{\mu_n}\, x + O\left(\frac{1}{\sqrt{\mu_n}}\right)$$

$$= \cos \pi nx + O\left(\frac{1}{n}\right).$$

Dividing by $\|y_2(\cdot, \mu_n)\|$,

$$g_n'(x) = \sqrt{2\mu_n}\cos \pi nx + O(1) = \sqrt{2}\,\pi n \cos \pi nx + O(1).$$

This proves the theorem. ∎

It will also be important to have asymptotic estimates of the squares of the eigenfunctions and the products

$$a_n(x, q) = y_1(x, \mu_n)y_2(x, \mu_n), \qquad n \ge 1.$$

Corollary 1. *For q in L^2,*

$$g_n^2 = 1 - \cos 2\pi nx + O\left(\frac{1}{n}\right)$$

$$\frac{d}{dx} g_n^2 = 2\pi n \sin 2\pi nx + O(1)$$

and

$$a_n = \frac{1}{2\pi n} \sin 2\pi nx + O\left(\frac{1}{n^2}\right)$$

$$\frac{d}{dx} a_n = \cos 2\pi nx + O\left(\frac{1}{n}\right).$$

All estimates hold uniformly on bounded subsets of $[0, 1] \times L^2$.

Proof.   The estimates of g_n^2 and $(d/dx)g_n^2$ follow from Theorem 4 by straightforward calculations. Also, $\sqrt{\mu_n} = n\pi + O(1/n)$ and the Basic Estimates for y_1 and y_2 yield

$$y_1(x, \mu_n) = \cos \pi nx + O\left(\frac{1}{n}\right)$$

$$y_2(x, \mu_n) = \frac{1}{\pi n} \sin \pi nx + O\left(\frac{1}{n^2}\right)$$

and

$$y_1'(x, \mu_n) = -\pi n \sin \pi nx + O(1)$$

$$y_2'(x, \mu_n) = \cos \pi nx + O\left(\frac{1}{n}\right)$$

uniformly on bounded subsets of $[0, 1] \times L^2$. Hence,

$$a_n = y_1 y_2 \bigg|_{\mu_n} = \frac{1}{2\pi n} \sin 2\pi nx + O\left(\frac{1}{n^2}\right)$$

and

$$\frac{d}{dx} a_n = (y_1 y_2' + y_1' y_2)\bigg|_{\mu_n}$$

$$= \cos^2 \pi nx - \sin^2 \pi nx + O\left(\frac{1}{n}\right)$$

$$= \cos 2\pi nx + O\left(\frac{1}{n}\right). \quad \blacksquare$$

Problem 3. (a) Show that the estimate of Theorem 4 can be improved to

$$\mu_n(q) = n^2\pi^2 + \int_0^1 q(t)\,dt - \langle \cos 2\pi nx, q \rangle + \ell_1^2(n).$$

(b) The function q is *odd* on $[0, 1]$, if $q(1 - x) = -q(x)$. Show that for odd q,

$$\mu_n(q) = n^2\pi^2 + \ell_1^2(n).$$

(c) Show that for q in the Sobolev space $H_{\mathbb{R}}^2$ with $\int_0^1 q(t)\,dt = 0$,

$$\mu_n(q) = n^2\pi^2 + \frac{1}{4n^2\pi^2}\left(q'(1) - q'(0) + \int_0^1 q^2(t)\,dt\right) + \ell_2^2(n).$$

The characteristic polynomial of an $n \times n$-matrix is equal to the product

$$\prod_{m=1}^{n}(\lambda_m - \lambda)$$

where $\lambda_1, \ldots, \lambda_n$ are the eigenvalues of the matrix. There is an analogous representation for $y_2(1, \lambda, q)$.

Theorem 5. *For q in L^2,*

$$y_2(1, \lambda, q) = \prod_{m \geq 1} \frac{\mu_m(q) - \lambda}{m^2\pi^2}.$$

Proof. We have $\mu_n = n^2\pi^2 + O(1)$ by Theorem 4. It follows from Lemma E.2 that the infinite product

$$p(\lambda) = \prod_{m \geq 1} \frac{\mu_m(q) - \lambda}{m^2\pi^2}$$

is an entire function of λ, which satisfies

$$p(\lambda) = \frac{\sin\sqrt{\lambda}}{\sqrt{\lambda}}\left(1 + O\left(\frac{\log n}{n}\right)\right)$$

uniformly on the circles $|\lambda| = r_n = (n + \frac{1}{2})^2\pi^2$ for n large enough. Its roots are precisely the Dirichlet eigenvalues of q, so the quotient $p(\lambda)/y_2(1, \lambda)$ is also an entire function. By the Basic Estimate for y_2 and Lemma 1,

$$y_2(1, \lambda) = \frac{\sin\sqrt{\lambda}}{\sqrt{\lambda}} + O\left(\frac{e^{|\mathrm{Im}\sqrt{\lambda}|}}{|\lambda|}\right)$$

$$= \frac{\sin\sqrt{\lambda}}{\sqrt{\lambda}}\left(1 + O\left(\frac{1}{n}\right)\right)$$

uniformly for $|\lambda| = r_n$. Hence,

$$\frac{p(\lambda)}{y_2(1, \lambda)} = 1 + O\left(\frac{\log n}{n}\right)$$

for $|\lambda| = r_n$, that is,

$$\sup_{|\lambda| = r_n} \left| \frac{p(\lambda)}{y_2(1, \lambda)} - 1 \right| \to 0$$

as $n \to \infty$. If follows from the maximum principle that the difference vanishes identically. Hence, $p(\lambda) = y_2(1, \lambda)$. ■

If the functions $y_2(1, \lambda, q)$ and $y_2(1, \lambda, p)$ are equal, then clearly $\mu_n(q) = \mu_n(p)$ for all $n \geq 1$. That is, q and p have the same Dirichlet spectrum. Now we know that the converse is also true. In other words, all the information in $y_2(1, \lambda)$ is already contained in the Dirichlet spectrum μ_n, $n \geq 1$.

Here are two useful consequences of the product formula.

Corollary 2.

(a)
$$\dot{y}_2(1, \mu_n) = \frac{-1}{n^2\pi^2} \prod_{m \neq n} \frac{\mu_m - \mu_n}{m^2\pi^2}$$

$$= \frac{(-1)^n}{2n^2\pi^2} \left(1 + O\left(\frac{\log n}{n}\right)\right)$$

(b)
$$\operatorname{sgn} \dot{y}_2(1, \mu_n) = (-1)^n = \operatorname{sgn} y_2'(1, \mu_n)$$

Proof. Part (a) follows from Lemma E.3. The first identity in (b) follows from (a), while the second one is a consequence of the first one and Theorem 2. ■

Problem 4. Prove Theorem 5, using Hadamard's theorem [Ti].
[Hint: Observe that $y_2(1, \lambda)$ is an entire function of order $\frac{1}{2}$, so that

$$y_2(1, \lambda) = c_0 \prod_{m \geq 1} \left(1 - \frac{\lambda}{\mu_m}\right)$$

$$= c_0 \prod_{m \geq 1} \frac{\mu_m - \lambda}{\mu_m}$$

$$= c_1 \prod_{m \geq 1} \frac{\mu_m - \lambda}{m^2\pi^2}$$

with

$$c_1 = c_0 \prod_{m \geq 1} \frac{m^2 \pi^2}{\mu_m}.$$

Now divide by

$$\frac{\sin \sqrt{\lambda}}{\sqrt{\lambda}} = \prod_{m \geq 1} \frac{m^2 \pi^2 - \lambda}{m^2 \pi^2}$$

and let λ tend to $-\infty$ to obtain $c_1 = 1$].

The rest of this chapter is devoted to properties of the eigenfunctions.

Theorem 6. (a) *For every q in L^2, the eigenfunction g_n, $n \geq 1$, has exactly $n + 1$ roots in $[0, 1]$. They are all simple.*
(b) *If q is even, then g_n is even when n is odd and odd when n is even.*

Even and odd mean even and odd about the point $\frac{1}{2}$.

The task of counting the roots of g_n is simplified by the following "deformation lemma":

Lemma 3. *Let h_t, $0 \leq t \leq 1$, be a family of real valued functions on $a \leq x \leq b$, which is jointly continuously differentiable in t and x. Suppose that for every t, h_t has a finite number of roots in $[a, b]$, all of which are simple, and has boundary values that are independent of t. Then h_0 and h_1 have the same number of roots in $[a, b]$.*

Intuitively speaking, the roots of h_t in the interior of $[a, b]$ move as t moves, but they can never collide or split, since they are all simple. Therefore, their number is independent of t.

Proof of Lemma 3. Suppose for convenience that

$$h_t(a) = 0 = h_t(b), \qquad 0 \leq t \leq 1.$$

Other cases are handled analogously. Fix t in $[0, 1]$. By assumption, the roots of h_t are simple. Therefore, we can place an interval around each of them, so that h_t changes sign once on the intervals in the interior, h_t' is bounded away from zero on all the intervals, and in addition, h_t is bounded away from zero on the complementary intervals. See Figure 3.

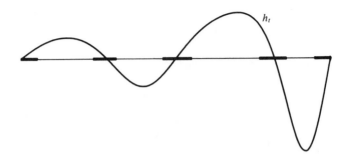

Figure 3.

By continuity, h_s and h'_s behave in exactly the same way on these intervals, if s is sufficiently close to t. Consequently, h_s has exactly the same number of roots as h_t, when s is close to t.

This argument applies to all t in $[0, 1]$, so the number of roots is independent of t. In particular, it is the same for h_0 and h_1. ∎

Proof of Theorem 6. (a) Since $g_n(x, q)$ does not vanish identically, its roots are all simple by Corollary 1.1. Their number is finite, since otherwise they would cluster at a multiple root.

Now consider the deformation h_t of g_n given by

$$h_t(x) = g_n(x, tq), \qquad 0 \le t, x \le 1.$$

Clearly, for each t, h_t vanishes at 0 and 1 and has only a finite number of roots in $[0, 1]$ all of which are simple. Thus we may apply Lemma 3, and it follows that $h_1 = g_n$ has as many roots in $[0, 1]$ as $h_0 = \sqrt{2} \sin \pi n x$. Hence, g_n has exactly $n + 1$ roots in $[0, 1]$.

(b) Let q be even. That is, $q(1 - x) = q(x)$ for $0 \le x \le 1$. Substituting $1 - x$ for x in

$$-g_n'' + q(x)g_n = \mu_n g_n,$$

we see that $g_n(1 - x)$ is another eigenfunction of q for μ_n with norm 1. Hence,

$$g_n(x) = -\operatorname{sgn} g_n'(1)g_n(1 - x).$$

Since $\operatorname{sgn} g_n'(1) = (-1)^n$ by Corollary 2, the claim follows. ∎

One of the main objectives of classical Sturm-Liouville theory is to generalize the Fourier sine expansion. The idea is to replace the sines by eigenfunctions of a boundary value problem for a second order equation.

This was a precursor of the abstract spectral theorem for self-adjoint operators. For the Dirichlet problem, the result is

Theorem 7. *For each q in L^2, the sequence*

$$g_n(x, q), \qquad n \geq 1$$

is an orthonormal basis for L^2.

Proof. Let $Q = -(d^2/dx^2) + q$. By partial integration and the boundary conditions on g_n and g_m,

$$(\mu_m - \mu_n)\langle g_m, g_n \rangle = \langle \mu_m g_m, g_n \rangle - \langle g_m, \mu_n g_n \rangle$$

$$= \langle Q g_m, g_n \rangle - \langle g_m, Q g_n \rangle$$

$$= [g_m, g_n] \Big|_0^1$$

$$= 0.$$

Since $\mu_m \neq \mu_n$ for $m \neq n$, this implies

$$\langle g_m, g_n \rangle = 0, \qquad m \neq n.$$

Since also $\langle g_n, g_n \rangle = 1$ by construction, the functions g_n, $n \geq 1$, are orthonormal.

It remains to show that this system is complete in L^2. We do this by comparing the vectors g_n with the vectors

$$e_n = \sqrt{2} \sin \pi n x, \qquad n \geq 1,$$

which do form an orthonormal basis of L^2.

Consider the linear operator A on L^2 defined by

$$Af = \sum_{n \geq 1} \langle f, e_n \rangle g_n.$$

A is well defined, since $\langle f, e_n \rangle$, $n \geq 1$, is an ℓ^2-sequence and g_n, $n \geq 1$, is an orthonormal basis. A is an isometry, since

$$\|Af\|^2 = \sum_{n \geq 1} |\langle f, e_n \rangle|^2 = \|f\|^2$$

by the orthogonality of the g_n and Parseval's identity. In particular, A is one-to-one. Also, by Theorem 4,

$$\sum_{n \geq 1} \|(A - I)e_n\|^2 = \sum_{n \geq 1} \|g_n - e_n\|^2$$

$$= \sum_{n \geq 1} O\left(\frac{1}{n^2}\right)$$

$$< \infty,$$

so $A - I$ is Hilbert-Schmidt and hence compact by Theorem D.1. It follows from the Fredholm Alternative, Theorem D.2, that A is onto L^2 and has a bounded inverse. Therefore, g_n, $n \geq 1$, is an orthonormal basis for L^2. ∎

There are more elementary proofs of the completeness of the Dirichlet eigenfunctions, which do not appeal directly to the Fredholm Alternative. See for instance Birkhoff and Rota [BR], who also make a number of interesting historical remarks. The method used here, however, is easy to generalize. Roughly speaking, any sequence of vectors, which is linearly independent and "sufficiently close" to some orthonormal basis, is itself a basis. Theorem D.3 makes this statement precise. It will be very useful in the sequel.

Actually, we will never need Theorem 7, but include it as a prototype for other completeness theorems. For us, it will be much more important to settle independence and completeness questions about *squares* of eigenfunctions, for example in the proof of Theorem 3.2. The next theorem is essential for this and many other purposes.

Recall that

$$a_n(x, q) = y_1(x, \mu_n)y_2(x, \mu_n), \qquad n \geq 1.$$

Theorem 8. *For $m, n \geq 1$,*

(a)
$$\left\langle g_m^2, \frac{d}{dx} g_n^2 \right\rangle = 0,$$

(b)
$$\left\langle a_m, \frac{d}{dx} g_n^2 \right\rangle = \tfrac{1}{2}\delta_{mn},$$

(c)
$$\left\langle a_m, \frac{d}{dx} a_n \right\rangle = 0.$$

It is only a slight exaggeration to say that Theorem 8 is the basis of almost everything else we are going to do.

Proof. (a) Integration by parts yields

$$\left\langle g_m^2, \frac{d}{dx} g_n^2 \right\rangle = \tfrac{1}{2} \int_0^1 (g_m^2 (g_n^2)' - (g_m^2)' g_n^2) \, dx$$

$$= \int_0^1 g_m g_n [g_m, g_n] \, dx.$$

This clearly vanishes for $m = n$. If $m \neq n$, then $\mu_m \neq \mu_n$, and we can use

$$[g_m, g_n]' = (\mu_m - \mu_n) g_m g_n$$

to obtain

$$\left\langle g_m^2, \frac{d}{dx} g_n^2 \right\rangle = \frac{1}{2} \frac{1}{\mu_m - \mu_n} [g_m, g_n]^2 \Big|_0^1 = 0.$$

(b) Again, by partial integration,

$$2 \left\langle a_m, \frac{d}{dx} g_n^2 \right\rangle = \int_0^1 (a_m (g_n^2)' - a_m' g_n^2) \, dx$$

$$= \int_0^1 (2 y_1 y_2 g_n g_n' - y_1' y_2 g_n^2 - y_1 y_2' g_n^2) \, dx$$

$$= \int_0^1 (y_2 g_n [y_1, g_n] + y_1 g_n [y_2, g_n]) \, dx,$$

where $y_j = y_j(x, \mu_m)$. If $m \neq n$, we may use

$$[y_j, g_n]' = (\mu_m - \mu_n) y_j g_n, \qquad j = 1, 2$$

to obtain

$$2 \left\langle a_m, \frac{d}{dx} g_n^2 \right\rangle = \frac{1}{\mu_m - \mu_n} [y_1, g_n][y_2, g_n] \Big|_0^1 = 0.$$

If $m = n$, then y_2 is a multiple of g_n, hence $[y_2, g_n] = 0$, and we get

$$2 \left\langle a_n, \frac{d}{dx} g_n^2 \right\rangle = \int_0^1 y_2 g_n [y_1, g_n] \, dx$$

$$= \int_0^1 g_n^2 [y_1, y_2] \, dx$$

$$= 1,$$

using the Wronskian identity.

(c) is proven in the same manner. ∎

For $q = 0$, the vectors 1,

$$g_n^2 - 1 = -\cos 2\pi nx,$$

$$\frac{d}{dx} g_n^2 = 2\pi n \sin 2\pi nx,$$

$n \geq 1$, are a basis for $L^2[0, 1]$. Theorem 8 and the asymptotic estimates of Corollary 1 make it possible to prove the same statement for every q.

First, let us specify the notions of linear independence and basis.

A sequence of vectors $v_1, v_2, \ldots$ in a Hilbert space is *linearly independent*, if none of the vectors v_n, $n \geq 1$, is contained in the closed linear span of all the other vectors v_m, $m \neq n$.

One might try to define a stronger notion of linear independence by requiring v_n to lie outside the weak closure of the span of v_m, $m \neq n$. However, if a sequence of vectors converges weakly, then the arithmetic means of some subsequence converge strongly to the same limit by the theorem of Banach-Saks [Ba]. Thus, the weak and strong closure of a linear space are the same, and nothing is gained.

A sequence of vectors $v_1, v_2, \ldots$ is a *basis* for a Hilbert space H, if there exists a Hilbert space $\hbar$ of sequences $\alpha = (\alpha_1, \alpha_2, \ldots)$ such that the correspondence

$$\alpha \to \sum_{n \geq 1} \alpha_n v_n$$

is a linear isomorphism between $\hbar$ and H.[4]

Theorem 9. *At every point in L^2, the vectors*

$$1, g_n^2 - 1, \qquad n \geq 1$$

are linearly independent. So are the vectors

$$\frac{d}{dx} g_n^2, \qquad n \geq 1.$$

The two sequences are mutually perpendicular, and together are a basis for L^2. Precisely,

$$(\xi, \eta) \to \sum_{n \geq 1} \xi_n \frac{d}{dx} g_n^2 + \eta_0 + \sum_{n \geq 1} \eta_n(g_n^2 - 1)$$

is a linear isomorphism between $\ell_1^2 \times \mathbb{R} \times \ell^2$ and L^2.

---

[4] A linear isomorphism between Banach spaces is an isomorphism between vector spaces, which is continuous and has a continuous inverse.

It follows from Theorem 6 that at an even point in L^2, the vectors $1, g_n^2 - 1$, $n \geq 1$, are a basis for the subspace of even functions in L^2, and the vectors $(d/dx)g_n^2$, $n \geq 1$, are a basis for the subspace of odd functions.

Proof. The vector $g_n^2 - 1$, $n \geq 1$, is not in the closed linear span of the vectors $1, g_m^2 - 1$, $m \neq n$, since

$$\left\langle g_n^2 - 1, \frac{d}{dx} a_n \right\rangle = \left\langle g_n^2, \frac{d}{dx} a_n \right\rangle = -\tfrac{1}{2},$$

but

$$\left\langle 1, \frac{d}{dx} a_n \right\rangle = 0,$$

$$\left\langle g_m^2 - 1, \frac{d}{dx} a_n \right\rangle = \left\langle g_m^2, \frac{d}{dx} a_n \right\rangle = 0, \qquad m \neq n$$

by Theorem 8. Similar arguments apply to all other vectors. The two sequences are mutually perpendicular, since

$$\left\langle 1, \frac{d}{dx} g_m^2 \right\rangle = 0, \qquad \left\langle g_n^2 - 1, \frac{d}{dx} g_m^2 \right\rangle = 0$$

for all m, n again by Theorem 8.

By Corollary 1,

$$g_n^2 - 1 = -\cos 2\pi nx + O\left(\frac{1}{n}\right)$$

$$1 = 1$$

$$\frac{1}{2\pi n} \frac{d}{dx} g_n^2 = \sin 2\pi nx + O\left(\frac{1}{n}\right).$$

We have shown that the vectors on the left are linearly independent. The vectors on the right without the error terms are (up to an irrelevant factor $\sqrt{2}$) an orthonormal basis of L^2, and the error terms are square summable. Therefore, Theorem D.3 applies. It follows that the vectors on the left are a basis with coefficients in $\ell^2 \times \mathbb{R} \times \ell^2$. From this, the last statement of the theorem follows. ∎

It is useful to write Theorem 9 in a slightly different form.

Corollary 3. *At every point q in L^2, the subspaces*

$$\left\{ \sum_{n \geq 1} \xi_n \frac{d}{dx} g_n^2 : \xi \in \ell_1^2 \right\},$$

$$\left\{ \eta_0 + \sum_{n \geq 1} \eta_n(g_n^2 - 1) : \eta \in \mathbb{R} \times \ell^2 \right\}$$

are perpendicular and closed. Their direct sum is all of L^2. In particular, they are the odd and even subspaces respectively when q is even.

Problem 5. (a) Show that

$$\left\| 1 - \frac{1}{N} \sum_{n=1}^{N} g_n^2 \right\| = O\left(\frac{\log N}{N}\right).$$

Conclude that 1 is in the closed linear span of g_n^2, $n \geq 1$.

  (b) Show that the vectors g_n^2, $(d/dx)g_n^2$, $n \geq 1$, span L^2 (that is, finite linear combinations are dense) and are linearly independent, but not a basis.

  The basic facts about the Dirichlet eigenvalues and eigenfunctions are now proven. We are ready to approach the inverse Dirichlet problem.

3 The Inverse Dirichlet Problem

In the last chapter we introduced the Dirichlet spectrum associated to a function q in L^2 and derived some of its properties. We found that the Dirichlet eigenvalues μ_n, $n \geq 1$, form a strictly increasing sequence of real numbers satisfying

$$\mu_n = n^2\pi^2 + c + \ell^2(n),$$

where $c = \int_0^1 q(t)\,dt$.

It is natural to ask whether these conditions actually characterize all possible Dirichlet spectra. For instance, the sequence $n^2\pi^2$, $n \geq 1$, is the Dirichlet spectrum for $q = 0$. Suppose the first eigenvalue π^2 is replaced by any number μ_1 below the second eigenvalue $4\pi^2$. Is the modified sequence still the Dirichlet spectrum of some q in L^2? Perhaps, μ_1 has to be chosen in a special way?

It is also natural to ask to what extent a point p in L^2 is determined by its Dirichlet spectrum. For instance, are there any functions q in L^2 besides $q = 0$ with Dirichlet eigenvalues $\mu_n = n^2\pi^2$, $n \geq 1$? More ambitiously, what does the *isospectral set*

$$M(p) = \{q \in L^2 : \mu_n(q) = \mu_n(p),\ n \geq 1\}$$

of all functions q with the same Dirichlet spectrum as p look like?

49

We have come to the inverse Dirichlet problem. It has two parts. First, describe the isospectral sets $M(p)$ for all p in L^2. Second, characterize all sequences of real numbers which arise as the Dirichlet spectrum of some q in L^2. These problems will occupy us in this and the next three chapters. Surprisingly, they both have complete solutions.

Problem 1.   Calculate the spectrum of the eigenvalue problem

$$iy' + q(x)y = \lambda y, \qquad 0 \le x \le 1$$

$$y(0) = y(1)$$

for q in L^2. Make a list of all possible spectra and decide when two qs have the same spectrum.

We begin by reformulating the inverse Dirichlet problem in more geometrical terms. For this purpose it is convenient to introduce the space S of all real, strictly increasing sequences $\sigma = (\sigma_1, \sigma_2, \ldots)$ of the form

$$\sigma_n = n^2\pi^2 + s + \tilde\sigma_n, \qquad n \ge 1$$

where $s \in \mathbb{R}$ and $\tilde\sigma = (\tilde\sigma_1, \tilde\sigma_2, \ldots) \in \ell^2$. By Theorem 2.4, the map

$$q \to \mu(q) = (\mu_1(q), \mu_2(q), \ldots)$$

from q to its sequence of Dirichlet eigenvalues sends L^2 into S. Characterizing spectra is equivalent to determining the image of μ, and isospectral sets are the fibers of μ, that is,

$$M(p) = \mu^{-1}(\mu(p))$$

$$= \{q \in L^2 : \mu(q) = \mu(p)\}.$$

Thus, solving the inverse Dirichlet problem amounts to analyzing the map μ.

For this purpose, observe that the correspondence between σ and $(s, \tilde\sigma)$ is a one-to-one map between S and an open subset of $\mathbb{R} \times \ell^2$. We shall *identify* S with this open subset, and refer to $(s, \tilde\sigma)$ as the standard coordinate system on S. This identification allows us to do analysis on S as if it were an open subset of a Banach space.[1]

In these coordinates the map μ becomes

$$q \to ([q], \tilde\mu(q)),$$

---

[1] See Appendix C, Example 3.

where

$$[q] = \int_0^1 q(t)\, dt$$

is the mean value of q, and $\tilde{\mu}(q) = (\tilde{\mu}_n(q), n \geq 1)$ is the sequence defined by

$$\tilde{\mu}_n(q) = \mu_n(q) - n^2\pi^2 - [q].$$

Theorem 1. *μ is a real analytic map from L^2 into S. Its derivative at q is the linear map from L^2 into $\mathbb{R} \times \ell^2$ given by*

$$d_q\mu(v) = ([v], \langle g_n^2 - 1, v \rangle, n \geq 1).$$

Proof. Fix p in L^2, and let N be an integer. By the analyticity of the Dirichlet eigenvalues, there exists a complex neighborhood $U_N \subset L_\mathbb{C}^2$ of p such that $\mu_1, \ldots, \mu_N$ extend to analytic functions on U_N and remain simple roots of $y_2(1, \lambda)$. We show that such a neighborhood can be chosen independently of N. That is, all eigenvalues extend analytically to some fixed complex neighborhood of p.

Choose $N > 2e^{\|p\|}$, and let $V \subset L_\mathbb{C}^2$ be a complex neighborhood of p such that

$$2e^{\|q\|} < N, \qquad q \in V.$$

By the Counting Lemma, there is a unique, *simple* root $\hat{\mu}_n(q)$ of $y_2(1, \lambda, q)$ in the egg-shaped region $|\sqrt{\lambda} - n\pi| < \pi/2$ for all $n > N$ and $q \in V$. Moreover, for real q,

$$\hat{\mu}_n(q) = \mu_n(q), \qquad q \in V \cap L_\mathbb{R}^2.$$

It follows from the implicit function theorem that $\hat{\mu}_n$ is an analytic extension of μ_n to V. Thus, all eigenvalues $\mu_1, \mu_2, \ldots$ extend analytically to the complex neighborhood $U = U_N \cap V$ of p, remaining simple roots of $y_2(1, \lambda)$.

There is a similar analytic extension of

$$g_n^2(x, q) = \frac{y_2^2(x, \mu_n)}{\dot{y}_2(1, \mu_n)y_2'(1, \mu_n)}, \qquad n \geq 1$$

to U, since $\dot{y}_2(1, \mu_n) \neq 0$ and $y_2'(1, \mu_n) \neq 0$ on U by construction. Proceeding word for word as in the proof of Theorem 2.4, one finds

$$\mu_n(q) = n^2\pi^2 + [q] - \langle \cos 2\pi n x, q \rangle + O\left(\frac{1}{n}\right)$$

uniformly on U. Hence, the extension of the map $\mu(q) = ([q], \tilde{\mu}(q))$ is a bounded map from U into the complexification of $\mathbb{R} \times \ell^2$, all of whose coefficients are analytic. By Theorem A.3, the map μ itself is analytic on U. In addition, its derivative is given by the derivatives of its components, and we have

$$\frac{\partial \tilde{\mu}_n}{\partial q} = g_n^2 - 1$$

by Theorem 2.3.

The preceding argument applies to any point p in L^2. Hence, μ is real analytic on L^2. ∎

It is easy to see that the map μ is not globally one-to-one. Let * denote the reflection across the subspace of all even functions in L^2. That is,

$$q^*(x) = q(1 - x), \qquad 0 \le x \le 1.$$

Lemma 1. $\mu(q^*) = \mu(q)$.

Loosely speaking, the Dirichlet spectrum cannot distinguish left from right.

Proof. One easily verifies that the reflected eigenfunction g_n^*, $n \ge 1$, is a Dirichlet eigenfunction of q^* with eigenvalue $\mu_n(q)$. It has exactly $n + 1$ roots in $[0, 1]$. Therefore,

$$\mu_n(q) = \mu_n(q^*), \qquad n \ge 1$$

by Theorem 2.6. ∎

A small, but enlightening step towards understanding the behavior of μ is to calculate its derivative at $q = 0$. It is the linear map from L^2 into $\mathbb{R} \times \ell^2$ given by

$$v \to ([v], -\langle \cos 2\pi nx, v \rangle, n \ge 1),$$

since at $q = 0$,

$$\frac{\partial \tilde{\mu}_n}{\partial q} = g_n^2(x, 0) - 1$$

$$= 2 \sin^2 \pi nx - 1$$

$$= -\cos 2\pi nx.$$

The kernel of $d_0\mu$ is the subspace of all functions in L^2 whose Fourier cosine coefficients vanish. Hence, it is the space

$$U = \{q \in L^2: q^* = -q\}$$

of all odd functions. Its orthogonal complement is the space

$$E = \{q \in L^2: q^* = q\}$$

of all even functions, and $d_0\mu$ is a linear isomorphism, in fact an isometry, between E and $\mathbb{R} \times \ell^2$ by the elementary theory of Fourier series.

To exploit this observation, let μ_E be the restriction of μ to E. By the last paragraph,

$$d_0\mu_E = (d_0\mu)\Big|_E$$

is a boundedly invertible map between E and $\mathbb{R} \times \ell^2$. It follows from the inverse function theorem that μ_E is a real analytic isomorphism between a neighborhood of 0 in E and a neighborhood of $\mu(0) = (n^2\pi^2, n \geq 1)$ in S. Hence, any small but arbitrary perturbation of the sequence $n^2\pi^2$, $n \geq 1$, is the Dirichlet spectrum of an even function. In particular,

$$\mu_1, 4\pi^2, 9\pi^2, \ldots$$

is a sequence of Dirichlet eigenvalues, whenever μ_1 is sufficiently close to π^2.

This analysis gives a local solution to one of the questions raised at the beginning of this chapter. Still, we want to know what happens when μ_1 is moved far to the left of π^2 or very close to $4\pi^2$. Such global questions will be answered in Chapter 6.

We can extract some more information. Using the splitting $L^2 = E \oplus U$, write

$$\mu(q) = \mu(e, u), \qquad q = e + u.$$

The derivative of this map with respect to e at $(0, 0)$ is an isometry of E and $\mathbb{R} \times \ell^2$. By the implicit function theorem, there exists a real analytic map $u \to e(u)$ between a neighborhood of 0 in U and a neighborhood of 0 in E such that

$$\mu(e(u), u) = \mu(0), \qquad e(0) = 0.$$

The graph of e in $E \oplus U$ is contained in $M(0)$ and is homeomorphic to its projection onto U. Consequently, there is an infinite dimensional set of functions with the same Dirichlet spectrum as $q = 0$.

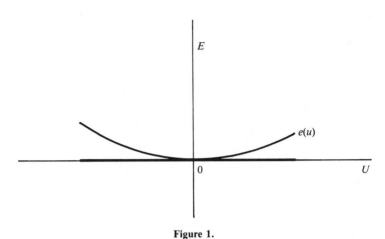

Figure 1.

Problem 2. (a) Show that the map

$$q \rightarrow (\mu(q), \tfrac{1}{2}(q - q^*))$$

from L^2 to $S \times U$ is a coordinate system in a neighborhood of $q = 0$.
[Hint: Write $L^2 = E \oplus U$. The map becomes $(e, u) \rightarrow (\mu(e, u), u)$.
Compute the derivative at 0 and use the inverse function theorem.]

(b) Show that in this coordinate system, the map from q to its Dirichlet spectrum is the projection of $S \times U$ onto S.

(c) Conclude from (a) and (b) that for each e in a sufficiently small neighborhood of 0 in E, the space $M(e)$ contains a set homeomorphic to an open subset of U.

Problem 3. The restriction of μ to the space U of odd functions is even:

$$\mu(q) = \mu(-q), \qquad q \in U.$$

Show that on U the Taylor expansion of μ_n at 0 is given by

$$\mu_n(q) = n^2\pi^2 + 2(-1)^{n+1}n^2\pi^2 \cdot S_2(1, n^2\pi^2, q) + O(\|q\|^4),$$

where S_2 is the second order term in the power series expansion of y_2 given in Theorem 1.1.
[Hint: By evenness,

$$\mu_n(q) = n^2\pi^2 + m_2(q) + O(\|q\|^4)]$$

Theorem 2. *μ_E is a local real analytic isomorphism at every point in E.*

Proof.   We show that the derivative of μ_E is boundedly invertible every-where on E. Then the result follows from the inverse function theorem.

In the standard coordinates on S, the derivative of μ_E is the linear map from E into $\mathbb{R} \times \ell^2$ given by

$$v \to ([v], \langle g_n^2 - 1, v \rangle, n \geq 1) = (\langle 1, v \rangle, \langle g_n^2 - 1, v \rangle, n \geq 1).$$

At every point in E, the vectors

$$1, g_n^2 - 1, \qquad n \geq 1$$

are a basis for E by Theorem 2.9. Therefore, $d_q \mu_E$ is boundedly invertible at every point in E.   ∎

Each eigenvalue μ_n, $n \geq 1$, is a compact function on L^2 by Theorem 2.3. On the other hand, μ_E is open by the preceding theorem and therefore *not* compact.

It turns out that μ_E is one-to-one not only locally, but even globally. In other words, an even function is uniquely determined by its Dirichlet spectrum. This important fact was first proved by Borg [Bo]. A simplified proof was given by Levinson [Le]. We give another proof.

Theorem 3. μ_E *is one-to-one on* E.

Before presenting the proof, we make a slight digression into complex analysis.

Lemma 2.   *Let f be a meromorphic function in the plane. If*

$$\sup_{|\lambda| = r_n} |f(\lambda)| = o\left(\frac{1}{r_n}\right)$$

for an unbounded sequence of positive real numbers r_n, then the sum of the residues of f is zero. In particular, if the residues of f are all nonnegative, then they are all zero.

The residues are not necessarily absolutely summable, so their "sum" may depend on the order in which they are added together. We define the sum of the residues, by the Cauchy residue theorem, as the limit of the partial sums

$$\frac{1}{2\pi i} \int_{|\lambda| = r_n} f(\lambda) \, d\lambda.$$

Proof of Lemma 2.   The sum of the residues of f is, by definition,

$$\lim_{n \to \infty} \frac{1}{2\pi i} \int_{|\lambda| = r_n} f(\lambda) \, d\lambda.$$

But

$$\left| \int_{|\lambda| = r_n} f(\lambda) \, d\lambda \right| \leq \int_{|\lambda| = r_n} |f(\lambda)| |d\lambda|$$

$$= o\left(\frac{1}{r_n}\right) \int_{|\lambda| = r_n} |d\lambda|$$

$$= o(1),$$

so the sum is zero.   ∎

Lemma 2 is an analogue of the fact that the sum of the residues of a meromorphic 1-form on a compact Riemann surface is zero. Since the plane is not compact, a growth condition is necessary. For instance, the residues of $\sqrt{\lambda}/\sin \sqrt{\lambda}$ are $2(-1)^n n^2 \pi^2$, $n \geq 1$, and their sum is not even defined.

Proof of Theorem 3.   Suppose q and p are even functions with $\mu(q) = \mu(p)$. We have to show that $q = p$.

Consider the meromorphic function

$$- \frac{[(y_2(x, \lambda, q) - y_2(x, \lambda, p)][(y_2(1 - x, \lambda, q) - y_2(1 - x, \lambda, p)]}{y_2(1, \lambda, q)}.$$

It has simple poles at μ_n, $n \geq 1$, otherwise it is regular. As a consequence of Theorem 2.6,

$$y_2(x, \mu_n) = (-1)^{n+1} y_2(1 - x, \mu_n), \qquad n \geq 1,$$

for q and p. It follows that the residue of our function at μ_n is

$$\frac{[y_2(x, \mu_n, q) - y_2(x, \mu_n, p)]^2}{(-1)^n \dot{y}_2(1, \mu_n, q)} \geq 0$$

using (b) of Corollary 2.2.

We now show that our function satisfies the hypothesis of Lemma 2 for $r_n = (n + \frac{1}{2})^2 \pi^2$. By Theorem 1.3, its numerator is bounded from above by a constant multiple of

$$\frac{e^{|\text{Im } \sqrt{\lambda}|x}}{|\lambda|} \frac{e^{|\text{Im } \sqrt{\lambda}|(1 - x)}}{|\lambda|} = \frac{e^{|\text{Im } \sqrt{\lambda}|}}{|\lambda|^2}.$$

To bound the denominator from below we observe that

$$\left| y_2(1, \lambda, q) - \frac{\sin \sqrt{\lambda}}{\sqrt{\lambda}} \right| \leq \frac{e^{\|q\|}}{|\sqrt{\lambda}|} \frac{e^{|\operatorname{Im} \sqrt{\lambda}|}}{|\sqrt{\lambda}|} \leq \frac{1}{8} \frac{e^{|\operatorname{Im} \sqrt{\lambda}|}}{|\sqrt{\lambda}|}$$

for $|\sqrt{\lambda}| \geq 8e^{\|q\|}$. Therefore, by Lemma 2.1,

$$
\begin{aligned}
|y_2(1, \lambda, q)| &\geq \left| \frac{\sin \sqrt{\lambda}}{\sqrt{\lambda}} \right| - \left| y_2(1, \lambda, q) - \frac{\sin \sqrt{\lambda}}{\sqrt{\lambda}} \right| \\
&\geq \frac{1}{4} \frac{e^{|\operatorname{Im} \sqrt{\lambda}|}}{|\sqrt{\lambda}|} - \frac{1}{8} \frac{e^{|\operatorname{Im} \sqrt{\lambda}|}}{|\sqrt{\lambda}|} \\
&\geq \frac{1}{8} \frac{e^{|\operatorname{Im} \sqrt{\lambda}|}}{|\sqrt{\lambda}|}
\end{aligned}
$$

for $|\lambda| = r_n$ and n sufficiently large. The quotient of the two bounds is $O(1/r_n^{3/2}) = o(1/r_n)$ as required.

So Lemma 2 applies, and it follows that

$$y_2(x, \mu_n, q) = y_2(x, \mu_n, p), \qquad n \geq 1.$$

Hence, q and p not only have the same Dirichlet spectrum, but also the same eigenfunctions. This is actually more than we need to know. If only, say, $y_2(x, \mu_1, q) = y_2(x, \mu_1, p)$, then

$$qy_2 = \mu_1(q)y_2 + y_2'' = \mu_1(p)y_2 + y_2'' = py_2$$

by the differential equation and therefore $q = p$ almost everywhere. ∎

By the last two theorems, μ_E is a one-to-one real analytic map from E into S, whose inverse is also real analytic. Thus, μ_E is a real analytic isomorphism between E and an open subset of S. We will see in Chapter 6 that μ_E maps onto S.

Problem 4. (Another proof of Theorem 3):
(a) Let z_1 and z_2 be the solutions of

$$-y'' + q(x)y = \lambda y, \qquad 0 \leq x \leq 1,$$

satisfying

$$z_1(1) = z_2'(1) = 1$$

$$z_1'(1) = z_2(1) = 0.$$

Show that

$$z_1(x, \lambda, q) = y_1(1 - x, \lambda, q^*)$$

$$z_2(x, \lambda, q) = -y_2(1 - x, \lambda, q^*).$$

(b)   Suppose q and p are even functions with the same Dirichlet spectrum. Show that

$$\begin{bmatrix} y_2(q) & z_2(q) \\ y_2'(q) & z_2'(q) \end{bmatrix} \begin{bmatrix} y_2(p) & z_2(p) \\ y_2'(p) & z_2'(p) \end{bmatrix}^{-1} = \begin{bmatrix} 1 & 0 \\ 0 & 1 \end{bmatrix},$$

where x and λ have been suppressed.

[Hint: Write down the inverse and multiply out. Check that the entries of the result are entire functions of λ bounded on circles of radius $(n + \frac{1}{2})^2\pi^2$. Then use the maximum principle.]

(c)   Conclude from (b) that two even functions with the same Dirichlet spectrum are equal.

Problem 5.   (Still another proof of Theorem 3 following Borg [Bo]):

(a)   Suppose q and p have the same Dirichlet spectrum. Show that[2]

$$\langle q - p, g_n(q)g_n(p) \rangle = 0, \qquad n \geq 1.$$

[Hint: Cross multiply the differential equations for $g_n(q)$, $g_n(p)$ and then subtract and integrate.]

(b)   Suppose q and p are even. Show that the sequence $1, g_n(q)g_n(p) - 1$, $n \geq 1$ is a basis for E.

[Hint: To prove independence, calculate the inner product between these vectors and

$$\frac{d}{dx} y_1(x, \mu_n(q), q)y_2(x, \mu_n(p), p), \qquad n \geq 1$$

as in the proof of Theorem 2.8. Otherwise imitate the proof of Theorem 2.9.]

(c)   Conclude from (a) and (b) that two even functions with the same Dirichlet spectrum are equal.

[Hint: $\langle q - p, 1 \rangle = 0$ by the asymptotics of the μ_n.]

The map from q to its Dirichlet spectrum is one-to-one on E, but, as we have seen, not on all of L^2. Additional data are necessary to determine q.

---

[2] We frequently suppress x and write $g_n(q)$ for $g_n(x, q)$ to simplify notation.

We introduce the quantities

$$\kappa_n(q) = \log \left| \frac{g_n'(1, q)}{g_n'(0, q)} \right|, \qquad n \ge 1,$$

where the logarithm will be useful later on. Equivalently,

$$\kappa_n(q) = \log |y_2'(1, \mu_n)|$$

$$= \log (-1)^n y_2'(1, \mu_n)$$

by Corollary 2.2. The numbers $\kappa_1, \kappa_2, \ldots$ are essentially the "terminal velocities" mentioned at the beginning of Chapter 2.

The following theorem lists the basic properties of the κ_n.

Theorem 4. *Each κ_n, $n \ge 1$, is a compact, real analytic function on L^2 with asymptotic behavior*

$$\kappa_n(q) = \frac{1}{2\pi n} \langle \sin 2\pi nx, q \rangle + O\left(\frac{1}{n^2}\right)$$

$$= \ell_1^2(n).$$

Its gradient is

$$\frac{\partial \kappa_n}{\partial q(t)} = a_n(t, q) - [a_n]g_n^2(t, q)$$

$$= \frac{1}{2\pi n} \sin 2\pi nt + O\left(\frac{1}{n^2}\right).$$

The error terms are uniform on bounded subsets of $[0, 1] \times L^2$.

Recall that $a_n = y_1(x, \mu_n)y_2(x, \mu_n)$.

Proof. $y_2'(1, \lambda, q)$ and $\mu_n(q)$ are compact, real analytic functions of (λ, q) and q, respectively. Since $y_2'(1, \mu_n(q), q)$ never vanishes,

$$\kappa_n(q) = \log (-1)^n y_2'(1, \mu_n(q), q)$$

is also compact and real analytic.

By the chain rule,

$$\frac{\partial \kappa_n}{\partial q(t)} = \frac{1}{y_2'(1)} \left(\frac{\partial}{\partial \lambda} y_2'(1) \frac{\partial \mu_n}{\partial q(t)} + \frac{\partial}{\partial q(t)} y_2'(1) \right) \Bigg|_{\lambda = \mu_n}.$$

Applying Theorem 1.6,

$$\frac{\partial}{\partial q(t)} y_2'(1) = y_2'(1)y_1(t)y_2(t) - y_1'(1)y_2^2(t)$$

and

$$\frac{\partial}{\partial \lambda} y_2'(1) = -y_2'(1)\int_0^1 y_1 y_2 \, dt + y_1'(1)\int_0^1 y_2^2 \, dt.$$

Inserting these identities into the expression for $\partial \kappa_n/\partial q$ and using

$$\frac{\partial \mu_n}{\partial q(t)} = g_n^2(t) = \frac{y_2^2(t, \mu_n)}{[y_2^2]},$$

one obtains

$$\frac{\partial \kappa_n}{\partial q(t)} = y_1(t, \mu_n)y_2(t, \mu_n) - \left(\int_0^1 y_1(t, \mu_n)y_2(t, \mu_n) \, dt\right)g_n^2(t)$$

$$= a_n(t) - [a_n]g_n^2(t)$$

$$= \frac{1}{2\pi n}\sin 2\pi nt + O\left(\frac{1}{n^2}\right).$$

The last line follows with Corollary 2.1.

Finally, since $\kappa_n(0) = 0$,

$$\kappa_n(q) = \int_0^1 \frac{d}{dt}\kappa_n(tq) \, dt$$

$$= \int_0^1 \left\langle \frac{1}{2\pi n}\sin 2\pi nt + O\left(\frac{1}{n^2}\right), q \right\rangle dt$$

$$= \frac{1}{2\pi n}\langle \sin 2\pi nx, q \rangle + O\left(\frac{1}{n^2}\right). \qquad \blacksquare$$

Problem 6. (a)   Show that for q in L^2, actually

$$\kappa_n(q) = \frac{1}{2\pi n}\langle \sin 2\pi nx, q \rangle + \ell_2^2(n).$$

(b)   Show that q is in the Sobolev space H^1 if and only if

$$\langle \cos 2\pi nx, q \rangle = \ell_1^2(n)$$

$$\langle \sin 2\pi nx, q \rangle = \frac{c}{n} + \ell_1^2(n),$$

where $c = (q(0) - q(1))/2\pi$.

[Hint: The Fourier sine coefficients of the odd function $x - \frac{1}{2}$ are $-\sqrt{2}/n\pi$, $n \geq 1$.]

(c) Show that

$$\mu_n(q) = n^2\pi^2 + [q] + \ell_1^2(n)$$

$$\kappa_n(q) = \frac{c}{n^2} + \ell_2^2(n)$$

if and only if q is in H^1 and $c = (q(0) - q(1))/4\pi^2$.

[Hint: Recall Problem 2.3 and integrate by parts.]

The gradients of κ_n and μ_n satisfy a set of simple relations, which will be essential later and are a direct consequence of Theorem 2.8.

Lemma 3. *For $m, n \geq 1$,*

(a)
$$\left\langle \frac{\partial\kappa_m}{\partial q}, \frac{d}{dx}\frac{\partial\kappa_n}{\partial q} \right\rangle = 0$$

(b)
$$\left\langle \frac{\partial\kappa_m}{\partial q}, \frac{d}{dx}\frac{\partial\mu_n}{\partial q} \right\rangle = \frac{1}{2}\delta_{mn}$$

(c)
$$\left\langle \frac{\partial\mu_m}{\partial q}, \frac{d}{dx}\frac{\partial\mu_n}{\partial q} \right\rangle = 0.$$

Proof. For instance, by Theorem 2.8,

$$\left\langle \frac{\partial\kappa_m}{\partial q}, \frac{d}{dx}\frac{\partial\mu_n}{\partial q} \right\rangle = \left\langle a_m - [a_m]g_m^2, \frac{d}{dx}g_n^2 \right\rangle$$

$$= \left\langle a_m, \frac{d}{dx}g_n^2 \right\rangle$$

$$= \frac{1}{2}\delta_{mn}. \qquad \blacksquare$$

By Theorem 4, the map

$$q \rightarrow \kappa(q) = (\kappa_1(q), \kappa_2(q), \ldots)$$

from q to its sequence of κ-values maps L^2 into the Hilbert space ℓ_1^2 of all real sequences $\xi = (\xi_1, \xi_2, \ldots)$ satisfying $\sum_{n\geq 1} n^2\xi_n^2 < \infty$. Combining κ and μ, we obtain a map

$$q \rightarrow (\kappa \times \mu)(q) = (\kappa(q), \mu(q))$$

from L^2 into the product space $\ell_1^2 \times S$. This map will play a central role in solving the inverse Dirichlet problem.

As a first step we show that the κ_n are sufficient supplementary data to the Dirichlet eigenvalues to determine q uniquely.

Theorem 5. $\kappa \times \mu$ *is one-to-one on* L^2.

Proof. Suppose that $\kappa(q) = \kappa(p)$ and $\mu(q) = \mu(p)$. We have to show that $q = p$.

We vary the proof of Theorem 3 and consider the meromorphic function

$$-\frac{[y_2(x, \lambda, q) - y_2(x, \lambda, p)][y_2(1 - x, \lambda, q^*) - y_2(1 - x, \lambda, p^*)]}{y_2(1, \lambda, q)},$$

which has simple poles at μ_n, $n \geq 1$. Our assumption $\kappa(q) = \kappa(p)$ implies

$$y_2'(1, \mu_n, q) = y_2'(1, \mu_n, p).$$

Moreover, by reflecting x into $1 - x$,

$$y_2(1 - x, \mu_n, q^*) = -\frac{y_2(x, \mu_n, q)}{y_2'(1, \mu_n, q)},$$

and similarly for p, since both sides are solutions of $-y'' + q^*y = \mu_n y$ with the same initial data at $x = 1$. Therefore, the residue of our function at μ_n is

$$\frac{[y_2(x, \mu_n, q) - y_2(x, \mu_n, p)]^2}{\dot{y}_2(1, \mu_n, q)y_2'(1, \mu_n, p)} \geq 0.$$

We can also verify the hypothesis of Lemma 2 for $r_n = (n + \frac{1}{2})^2\pi^2$ by the same estimates as in the proof of Theorem 3. The argument is then completed as before. ■

While μ does not change upon reflecting q by Lemma 1, κ changes sign. From this and the preceding uniqueness result we obtain a characterization of the even functions.

Lemma 4.

$$\kappa(q^*) = -\kappa(q).$$

In particular, q is even if and only if $\kappa(q) = 0$.

Proof.   In the proof of Theorem 5 we noticed that

$$y_2(1 - x, \mu_n(q), q^*) = - \frac{y_2(x, \mu_n(q), q)}{y_2'(1, \mu_n(q), q)}.$$

On the left we can replace $\mu_n(q)$ by $\mu_n(q^*)$ by Lemma 1, hence

$$\kappa_n(q^*) = \log (-1)^n \frac{1}{y_2'(1, \mu_n(q), q)} = -\kappa_n(q).$$

This holds for all $n \geq 1$, so $\kappa(q^*) = -\kappa(q)$. In particular, if q is even, then $q = q^*$, and consequently $\kappa(q) = 0$.

Conversely, suppose $\kappa(q) = 0$. Then

$$\kappa(q^*) = -\kappa(q) = \kappa(q)$$

$$\mu(q^*) = \mu(q),$$

the second line being the content of Lemma 1. Since the map $\kappa \times \mu$ is one-to-one, we obtain $q^* = q$. That is, q is even.   ∎

Next we study the analytic properties of the map $\kappa \times \mu$. It is not very surprising that this map is a local real analytic isomorphism at every point. More remarkably, its derivative can be inverted explicitly.

It is convenient to introduce some notation. For $n \geq 1$, set

$$V_n(x, q) = 2 \frac{d}{dx} g_n^2 = 2 \frac{d}{dx} \frac{\partial \mu_n}{\partial q}$$

$$W_n(x, q) = -2 \frac{d}{dx} (a_n - [a_n]g_n^2) = -2 \frac{d}{dx} \frac{\partial \kappa_n}{\partial q}.$$

We have

$$V_n = 4\pi n \sin 2\pi n x + O(1)$$

$$W_n = -2 \cos 2\pi n x + O\left(\frac{1}{n}\right)$$

uniformly on bounded subsets of $[0, 1] \times L^2$ by Corollary 2.1.

Theorem 6. $\kappa \times \mu$ *is a local real analytic isomorphism at every point in* L^2. *Moreover, the inverse of* $d_q(\kappa \times \mu)$ *is the linear map from* $\ell_1^2 \times \mathbb{R} \times \ell^2$ *onto* L^2 *given by*

$$(d_q(\kappa \times \mu))^{-1}(\xi, \eta) = \sum_{n \geq 1} \xi_n V_n + \eta_0 + \sum_{n \geq 1} \eta_n W_n.$$

Proof. μ is real analytic on L^2 by Theorem 1. We have to show that also κ is real analytic.

Fix p in L^2. We showed in the proof of Theorem 1 that there exists a complex neighborhood U of p, to which all μ_n and g_n^2 extend as analytic functions of q, satisfying the estimates of Theorem 2.4. Therefore,

$$\kappa_n = \log(-1)^n y_2'(1, \mu_n), \qquad n \geq 1$$

and

$$\frac{\partial \kappa_n}{\partial q} = a_n - [a_n]g_n^2, \qquad n \geq 1$$

are also analytic on U, since $y_2'(1, \mu_n)$ does not vanish. Moreover, imitating the proof of Theorem 4, we have

$$\kappa_n = \frac{1}{2\pi n} \langle \sin 2\pi nx, q \rangle + O\left(\frac{1}{n^2}\right), \qquad n \geq 1$$

uniformly on U. It follows that κ extends to a bounded map from U into the complexification of ℓ_1^2, all of whose components are analytic. Hence, κ is analytic on U by Theorem A.3.

The preceding argument applies to any point p. Therefore, the map κ is analytic on L^2.

By Theorem A.3, the derivative of $\kappa \times \mu$, in the standard coordinates on S, is the linear map from L^2 into $\ell_1^2 \times \mathbb{R} \times \ell^2$ given by

$$v \to \left(\left\langle \frac{\partial \kappa_n}{\partial q}, v \right\rangle, n \geq 1; [v]; \left\langle \frac{\partial \tilde{\mu}_n}{\partial q}, v \right\rangle, n \geq 1 \right).$$

We show that this map is boundedly invertible. Then, by the inverse function theorem, $\kappa \times \mu$ is a local real analytic isomorphism.

By Theorem 4 and Corollary 2.1,

$$2\pi n \frac{\partial \kappa_n}{\partial q} = \sin 2\pi nx + O\left(\frac{1}{n}\right), \qquad n \geq 1$$

$$1 = 1$$

$$\frac{\partial \tilde{\mu}_n}{\partial q} = -\cos 2\pi nx + O\left(\frac{1}{n}\right), \qquad n \geq 1.$$

These vectors are linearly independent, since, by Lemma 3, for each one of them there is another vector which is perpendicular to all vectors but the given one. Moreover, the vectors on the right, without the error terms, are an orthonormal basis of L^2 (up to an irrelevant factor $\sqrt{2}$), and the error terms are square summable. Thus, Theorem D.3 applies, and the above map, with $\partial\kappa_n/\partial q$ replaced by $2\pi n \cdot \partial\kappa_n/\partial q$, is a linear isomorphism between L^2 and $\ell^2 \times \mathbb{R} \times \ell^2$. Hence, the original map is a linear isomorphism between L^2 and $\ell_1^2 \times \mathbb{R} \times \ell^2$.

Now consider the inverse. By the asymptotics of V_n and W_n,

$$u = \sum_{n \geq 1} \xi_n V_n + \eta_0 + \sum_{n \geq 1} \eta_n W_n$$

converges in L^2 when $\xi \in \ell_1^2$ and $\eta \in \mathbb{R} \times \ell^2$. By Lemma 3,

$$\left\langle \frac{\partial\kappa_n}{\partial q}, u \right\rangle = \xi_n$$

$$\langle 1, u \rangle = \eta_0$$

$$\left\langle \frac{\partial\tilde{\mu}_n}{\partial q}, u \right\rangle = \eta_n,$$

hence

$$d_q(\kappa \times \mu)(u) = (\xi, \eta).$$

This proves the theorem. ∎

As a special case we obtain the inverse of the derivative of μ_E. By Theorems 2 and 3, this map is a real analytic isomorphism between E and an open subset of S.

Corollary 1. *For q in E, the inverse of $d_q\mu_E$ is the linear map from $\mathbb{R} \times \ell^2 \simeq T_{\mu(q)}S$ onto $E \simeq T_qE$ given by*

$$(d_q\mu_E)^{-1}(\eta) = \eta_0 + \sum_{n \geq 1} \eta_n W_n.$$

In particular, the sequence $1, W_n, n \geq 1$, is a basis for E.

It is implicit that the functions W_n are even at an even point q.

Proof. By the last theorem, $\kappa \times \mu$ is a real analytic isomorphism between L^2 and an open subset of $\ell_1^2 \times \mathbb{R} \times \ell^2$. In particular, by Lemma 4, the linear subspace E is mapped isomorphically to the intersection of this open set with the linear space $0 \times \mathbb{R} \times \ell^2$. It follows that μ_E^{-1} is the restriction of $(\kappa \times \mu)^{-1}$ to this intersection, and $(d_q\mu_E)^{-1}$ is the restriction of $(d_q(\kappa \times \mu))^{-1}$ to the linear space $0 \times \mathbb{R} \times \ell^2$. This proves the identity.

The second statement is immediate, since $(d_q\mu_E)^{-1}$ is a linear isomorphism. ∎

4 Isospectral Sets

How does the isospectral set

$$M(p) = \mu^{-1}(\mu(p))$$

$$= \{q \in L^2 : \mu(q) = \mu(p)\}$$

lie inside L^2? Is it connected? Bounded? Is it a submanifold? In this chapter, we shall describe $M(p)$ by answering questions of this kind.

Our basic intuition about isospectral sets comes from a simple picture in which we view $M(p)$ as the intersection of infinitely many hypersurfaces. Namely,

$$M(p) = \bigcap_{n \geq 1} M_n(p),$$

where

$$M_n(p) = \{q \in L^2 : \mu_n(q) = \mu_n(p)\}$$

is the set of all functions in L^2 with the same nth Dirichlet eigenvalue as p.

Each of these sets is a real analytic submanifold of L^2 of codimension one, since the gradients

$$\frac{\partial \mu_n}{\partial q(x)} = g_n^2(x, q)$$

never vanish identically.

We saw at the end of Chapter 2 that the gradients g_n^2, $n \geq 1$, are linearly independent at every point in L^2. This result now tells us that every finite intersection

$$M_1(p) \cap \cdots \cap M_n(p)$$

is a real analytic submanifold of L^2, whose normal space at q is spanned by the vectors[1]

$$g_1^2(q), \ldots, g_n^2(q).$$

Hence there is reason to conjecture that the infinite intersection $M(p) = \bigcap_{n \geq 1} M_n(p)$ is also a real analytic submanifold of L^2, whose normal space at q is the closed linear span of the gradients $g_n^2(q)$, $n \geq 1$.

The linear independence of the gradients, however, is not enough to guarantee this. A simple counterexample is given in Appendix C. More has to be done. We will show that at every point q on $M(p)$, the derivative $d_q \mu$ is a linear isomorphism between the orthogonal complement of its kernel and the tangent space to S at $\mu(p)$. Then $\mu(p)$ is a regular value of the map μ, and by the regular value theorem in Appendix C, $M(p)$ is a real analytic submanifold of L^2. Moreover, its tangent space at q is

$$T_q M(p) = \ker d_q \mu,$$

the kernel of $d_q \mu$, and its normal space at q is

$$N_q M(p) = \ker^{\perp} d_q \mu,$$

its orthogonal complement. As we will see in a moment, they are the closed linear spans of the vectors $(d/dx)g_n^2$, $n \geq 1$, and g_n^2, $n \geq 1$, respectively.

Let us verify our intuitive picture. To simplify notation, set

$$U_0 = 1$$

$$U_n = g_n^2 - 1, \qquad n \geq 1,$$

---

[1] We frequently suppress x, and for example write $g_n(q)$ for $g_n(x, q)$ to simplify notation.

and, as in Chapter 3,

$$V_n = 2\frac{d}{dx}g_n^2, \qquad n \geq 1.$$

These vectors depend on x and q. We write

$$U_\eta = \sum_{n \geq 0} \eta_n U_n$$

and

$$V_\xi = \sum_{n \geq 1} \xi_n V_n,$$

where $\eta \in \mathbb{R} \times \ell^2$ and $\xi \in \ell_1^2$. By the asymptotics of $g_n^2 - 1$ and $(d/dx)g_n^2$, these sums converge in L^2. By Corollary 2.3,

$$\{U_\eta \colon \eta \in \mathbb{R} \times \ell^2\},$$

$$\{V_\xi \colon \xi \in \ell_1^2\}$$

are perpendicular, closed linear subspaces, whose direct sum is L^2.

Theorem 1. (a) *For all p in L^2, $M(p)$ is a real analytic submanifold of L^2 lying in the hyperplane of all functions with mean $[p]$.*
  (b) *At every point q in $M(p)$, the normal space is*

$$N_q M(p) = \{U_\eta(q) \colon \eta \in \mathbb{R} \times \ell^2\},$$

and the tangent space is

$$T_q M(p) = \{V_\xi(q) \colon \xi \in \ell_1^2\}.$$

Proof. Let q be a point in $M(p)$. The derivative of μ at q is the linear map from L^2 into $\mathbb{R} \times \ell^2$ given by

$$d_q\mu(w) = (\langle U_n, w \rangle, n \geq 0).$$

We have the splitting

$$L^2 = \ker_q \oplus \ker_q^\perp,$$

where $\ker_q$ is the kernel of $d_q\mu$. We show that the restriction of $d_q\mu$ to $\ker_q^\perp$ is boundedly invertible. Then, $\mu(p)$ is a regular value of μ, and by the regular value theorem, $M(p)$ is a real analytic submanifold of L^2. Moreover, $T_q M(p) = \ker_q$ and $N_q M(p) = \ker_q^\perp$.

We have

$$\ker_q^\perp = \left\{ U_\eta = \sum_{n \geq 0} \eta_n U_n \colon \eta \in \mathbb{R} \times \ell^2 \right\},$$

since, by the form of the derivative, the vectors U_n are perpendicular to $\ker_q$, any vector perpendicular to them belongs to $\ker_q$, and the right hand side is closed by Corollary 2.3. By the way, the sequence U_n, $n \geq 0$, is a basis of $\ker_q^\perp$.

With respect to this basis, the restriction of $d_q \mu$ to $\ker_q^\perp$ is represented by the matrix operator

$$D = (\langle U_i, U_j \rangle)_{i,j \geq 0},$$

acting on $\mathbb{R} \times \ell^2$. By Corollary 2.1,

$$\langle U_i, U_j \rangle - \delta_{ij} = \left\langle \cos 2\pi i x, O\left(\frac{1}{j}\right) \right\rangle + \left\langle \cos 2\pi j x, O\left(\frac{1}{i}\right) \right\rangle + O\left(\frac{1}{ij}\right).$$

Using the Bessel inequality, we have

$$\sum_{i,j \geq 0} \left| \left\langle \cos 2\pi i x, O\left(\frac{1}{j}\right) \right\rangle \right|^2 \leq \sum_{j \geq 0} \left| O\left(\frac{1}{j}\right) \right|^2 < \infty,$$

and similarly for the other terms. This shows that $D - I$ is Hilbert-Schmidt, hence compact. D is also one-to-one, since the U_n are a basis of $\ker_q^\perp$. It follows from the Fredholm alternative that D is boundedly invertible, as was to be proven.

We have shown that $M(p)$ is a real analytic submanifold with normal space

$$N_q M(p) = \ker_q^\perp = \{ U_\eta(q) \colon \eta \in \mathbb{R} \times \ell^2 \}.$$

Its tangent space is the orthogonal complement of $\ker_q^\perp$, that is,

$$T_q M(p) = \ker_q = \{ V_\xi(q) \colon \xi \in \ell_1^2 \}$$

by Corollary 2.3. Finally, for q in $M(p)$,

$$[q] = \lim_{n \to \infty} (\mu_n(q) - n^2 \pi^2)$$

$$= \lim_{n \to \infty} (\mu_n(p) - n^2 \pi^2) = [p]$$

by Theorem 2.4. So $M(p)$ lies in the hyperplane of all functions with mean $[p]$. ∎

Problem 1. Show that a tangent vector V can be expanded as

$$V = \sum_{n \geq 1} \langle V, a_n \rangle V_n,$$

and a normal vector U as

$$U = [U] - 2 \sum_{n \geq 1} \langle U, \frac{d}{dx} a_n \rangle U_n.$$

Problem 2. Show that for all real λ,

$$\frac{\partial}{\partial q} y_2(1, \lambda) \in N_q M(p),$$

$$\frac{d}{dx} \frac{\partial}{\partial q} y_2(1, \lambda) \in T_q M(p).$$

Verify the identities

$$\frac{\partial}{\partial q} y_2(1, \lambda) = -\dot{y}_2(1, \lambda) + \sum_{n \geq 1} \frac{y_2(1, \lambda)}{\mu_n - \lambda} U_n,$$

$$\frac{d}{dx} \frac{\partial}{\partial q} y_2(1, \lambda) = \frac{1}{2} \sum_{n \geq 1} \frac{y_2(1, \lambda)}{\mu_n - \lambda} V_n.$$

Our picture of intersecting hypersurfaces is primitive, but quickly led us to a good understanding of the local properties of $M(p)$. There is another, more global picture of isospectral sets: we view $M(p)$ as a horizontal slice of the real analytic coordinate system $\kappa \times \mu$ on L^2 constructed in Chapter 3. Namely, $M(p)$ is the set of all points whose μ-coordinate is $\mu(p)$.

We have, almost at once, an alternate version of Theorem 1.

Theorem 1*. (a) *For all p in L^2, $M(p)$ is a real analytic submanifold of L^2, lying in the hyperplane of all functions with mean $[p]$.*
  (b) *At every point q in $M(p)$, the normal space is*

$$N_q M(p) = \{U_\eta : \eta \in \mathbb{R} \times \ell^2\},$$

and the tangent space is

$$T_q M(p) = \{V_\xi : \xi \in \ell_1^2\}.$$

  (c) *κ is a global coordinate system on $M(p)$. Its derivative $d_q \kappa$ is an iso-morphism between $T_q M(p)$ and $T_{\kappa(q)} \ell_1^2 \simeq \ell_1^2$, which is given by*

$$d_q \kappa(V_\xi) = \xi.$$

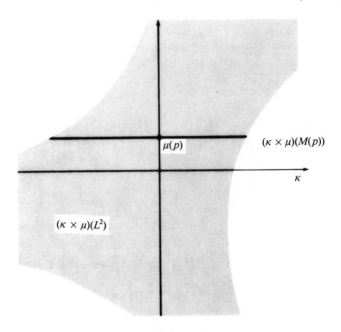

$\mu(p)$

$(\kappa \times \mu)(M(p))$

κ

$(\kappa \times \mu)(L^2)$

Figure 1.

In more detail, κ is a real analytic isomorphism between $M(p)$, as a submanifold of L^2, and its image $\kappa(M(p))$, an open subset of ℓ_1^2.

Proof. $M(p)$ is a real analytic submanifold, and κ is a global coordinate system on it, by the properties of the $\kappa \times \mu$-map and the definition of submanifold in Appendix C. Its tangent space is

$$T_q M(p) = \{(d_q(\kappa \times \mu))^{-1}(\xi, 0): \xi \in \ell_1^2\}$$

$$= \left\{ \sum_{n \geq 1} \xi_n V_n : \xi \in \ell_1^2 \right\}$$

by Theorem 3.6, and its orthogonal complement is

$$N_q M(p) = \left\{ \sum_{n \geq 0} \eta_n U_n : \eta \in \mathbb{R} \times \ell^2 \right\}$$

by Corollary 2.3. Finally,

$$d_q \kappa(V_\xi) = \xi$$

by Lemma 3.3. ∎

Every isospectral set is a real analytic submanifold of L^2 isomorphic to an open subset of ℓ_1^2. This more than answers one of the questions raised at the beginning of this chapter. We now go on to answer the other questions.

Let $M(p)$ be an isospectral submanifold, and let q be a point on it. Every tangent vector to $M(p)$ at q is of the form

$$V_\xi = \sum_{n \geq 1} \xi_n V_n$$

with uniquely determined coefficients $\xi = (\xi_1, \xi_2, \ldots)$ in ℓ_1^2. These coefficients in turn uniquely determine a tangent vector at every other point on $M(p)$ by the same expression, since the V_n are globally defined. In this way, every tangent vector V_ξ at a given point determines a globally defined vectorfield on $M(p)$, which we denote by the same symbol. The space of these vectorfields is isomorphic to any tangent space $T_q M(p)$ and hence isomorphic to ℓ_1^2.

We are going to study the solution curves of V_ξ. By definition, a curve

$$\phi^t(q) = \phi^t(q, V_\xi), \qquad a < t < b$$

on $M(p)$ is a solution curve of the vectorfield V_ξ with initial value q, if

$$\frac{d}{dt} \phi^t(q) = V_\xi(\phi^t(q)), \qquad a < t < b$$

and $\phi^0(q) = q$. We implicitly assume that the curve is differentiable in t, and that $a < 0 < b$. See Appendices B and C for solution curves of vectorfields in general.

The local existence and uniqueness of such solution curves is immediate. In the κ-coordinate system on $M(p)$, the vectorfield V_ξ becomes the constant vectorfield ξ by Theorem 1*. Consequently, any solution curve in this coordinate system is a straight line, and we must have

$$\kappa(\phi^t(q, V_\xi)) = \kappa(q) + t\xi.$$

Applying the real analytic map κ^{-1}, we obtain an expression for $\phi^t(q, V_\xi)$. It follows that $\phi^t(q, V_\xi)$ is an *analytic function of t, ξ and q*, and is defined as long as the straight line $\kappa(q) + t\xi$ remains inside the open set $\kappa(M(p))$.

The local existence and uniqueness of $\phi^t(q, V_\xi)$ may also be established without reference to the special coordinates κ, by looking at $M(p)$ alone. The argument is outlined in the following Problem.

Problem 3 (Another local existence and uniqueness proof). Show directly
that V_ξ is a real analytic vectorfield on $M(p)$, in fact, on all of L^2. Apply the
local existence and uniqueness theorem in Appendix C and conclude that
there is a unique real analytic solution curve of V_ξ for every initial value on
$M(p)$.[2]
[Hint: Fix ξ in ℓ_1^2. One can show as in the proof of Theorem 3.1 that

$$V_n = 4\pi n \sin 2\pi nx + O(1)$$

uniformly on a complex neighborhood of every point in L^2. Thus the sum
$\sum \xi_n V_n$ consists of two terms, one independent of q, and one converging
uniformly on every such neighborhood. Apply Theorem A.2 to conclude that
V_ξ is real analytic for each ξ].

Our immediate goal is to show that each solution curve exists for all time.
This is done in the usual manner by deriving an a priori bound for $\|\phi'(q, V_\xi)\|$.
First, two technical lemmas.

Lemma 1. *For $n \geq 1$,*

$$\langle q, V_n \rangle = 4\delta_n(q) \sinh \kappa_n(q),$$

where

$$\delta_n(q) = \frac{(-1)^n}{\dot{y}_2(1, \mu_n)}.$$

Proof. Using the differential equation and Theorem 2.2 we find

$$\left\langle q, \frac{d}{dx} g_n^2 \right\rangle = \int_0^1 2qg_n g_n' \, dx$$

$$= \int_0^1 2(g_n'' + \mu_n g_n)g_n' \, dx$$

$$= \int_0^1 \frac{d}{dx}((g_n')^2 + \mu_n g_n^2) \, dx$$

$$= ((g_n')^2 + \mu_n g_n^2)\Big|_0^1$$

[2] This is not enough to show, however, that $\phi'(q, V_\xi)$ is also real analytic in ξ. For the stronger
result, one must show that V_ξ is real analytic in q and ξ.

$$= \frac{(y_2'(x, \mu_n))^2}{\dot{y}_2(1, \mu_n) y_2'(1, \mu_n)} \Bigg|_0^1$$

$$= \frac{1}{\dot{y}_2(1, \mu_n)} \left(y_2'(1, \mu_n) - \frac{1}{y_2'(1, \mu_n)} \right).$$

Thus, by the definition of κ_n and δ_n,

$$\langle q, V_n \rangle = 2 \left\langle q, \frac{d}{dx} g_n^2 \right\rangle$$

$$= \frac{2(-1)^n}{\dot{y}_2(1, \mu_n)} (e^{\kappa_n} - e^{-\kappa_n})$$

$$= 4\delta_n \sinh \kappa_n. \qquad \blacksquare$$

We notice that the functions δ_n are positive on L^2 and have the asymptotic behavior

$$\delta_n = 2n^2\pi^2 \left(1 + O\left(\frac{\log n}{n}\right) \right)$$

by Corollary 2.2. Moreover, by the same Corollary, they are constant on every isospectral set, since

$$\dot{y}_2(1, \mu_n) = \frac{-1}{n^2\pi^2} \prod_{m \neq n} \frac{\mu_m - \mu_n}{m^2\pi^2}$$

depends only on the Dirichlet spectrum of q.

Lemma 2.

$$\|\phi^t(q, V_\xi)\|^2 = \|q\|^2 + 8 \sum_{n \geq 1} \delta_n(\cosh(\kappa_n + t\xi_n) - \cosh \kappa_n),$$

where δ_n and κ_n are evaluated at q.

Proof. For $\phi^t(q) = \phi^t(q, V_\xi)$, we have

$$\frac{1}{2} \frac{d}{ds} \|\phi^s(q)\|^2 \Bigg|_{s=0} = \langle q, V_\xi(q) \rangle$$

$$= \sum_{n \geq 1} \xi_n \langle q, V_n(q) \rangle$$

$$= 4 \sum_{n \geq 1} \xi_n \delta_n(q) \sinh \kappa_n(q)$$

by Lemma 1. To obtain the derivative at time $t \neq 0$, replace q in the above expression by $\phi^t(q)$ and make use of the fact that $\phi^{s+t}(q) = \phi^s(\phi^t(q))$. Then

$$\frac{1}{2}\frac{d}{dt}\|\phi^t(q)\|^2 = \frac{1}{2}\frac{d}{ds}\|\phi^{s+t}(q)\|^2\bigg|_{s=0}$$

$$= 4 \sum_{n \geq 1} \xi_n \delta_n(q) \sinh{(\kappa_n(q) + t\xi_n)},$$

since $\delta_n(\phi^t(q)) = \delta_n(q)$ and $\kappa_n(\phi^t(q)) = \kappa_n(q) + t\xi$. The right hand side converges uniformly on bounded intervals of time, because

$$\xi_n \delta_n \sinh{(\kappa_n + t\xi_n)} = O(\xi_n \delta_n(\kappa_n + t\xi_n))$$

$$= O(n^2 \xi_n^2)$$

for bounded t, and $\sum n^2 \xi_n^2 < \infty$. We may therefore integrate under the summation sign to obtain

$$\|\phi^t(q)\|^2 - \|q\|^2 = \int_0^t \frac{d}{ds}\|\phi^s(q)\|^2\,ds$$

$$\underline{} = 8 \sum_{n \geq 1} \delta_n \int_0^t \xi_n \sinh{(\kappa_n + s\xi_n)}\,ds$$

$$= 8 \sum_{n \geq 1} \delta_n(\cosh{(\kappa_n + t\xi_n)} - \cosh{\kappa_n}). \quad \blacksquare$$

Theorem 2.   *For every q on $M(p)$ and every ξ in ℓ_1^2, the solution curve $\phi^t(q, V_\xi)$ exists for all time.*

This together with the identity of Lemma 2 implies that the set $M(p)$ is unbounded.

Proof.   Fix q in $M = M(p)$ and ξ in ℓ_1^2. Theorem 2 is equivalent to the statement that the straight line $\kappa(q) + t\xi$ never leaves the open set $\kappa(M)$. Suppose, to the contrary, it does so for some positive t, say. Then there exists $t^* > 0$ such that $\kappa(q) + t^*\xi$ lies on the boundary of $\kappa(M)$, while the segment $\kappa(q) + t\xi$, $0 < t < t^*$, lies in its interior. By Lemma 2,

$$\sup_{0 < t < t^*} \|\phi^t(q)\| < \infty.$$

Hence we can choose a sequence t_n converging to t^* from below such that $\phi^{t_n}(q)$ converges weakly to some point q_*. This point lies on M with

$$\kappa(q_*) = \kappa(q) + t^*\xi,$$

since the μ_n and κ_n are compact functions on L^2. It follows that $\kappa(q) + t*\xi$ lies in the interior of $\kappa(M)$. That is a contradiction. ∎

Problem 4. Show that the flows of any two vectorfields V_ζ and V_ξ commute. That is,

$$\phi^s(\phi^t(q, V_\xi), V_\zeta) = \phi^t(\phi^s(q, V_\zeta), V_\xi)$$

for all s, t. Using the notation of Appendix B, this can be written more succinctly as

$$\Phi^s_\zeta \circ \Phi^t_\xi = \Phi^t_\xi \circ \Phi^s_\zeta,$$

where Φ_ζ, Φ_ξ denote the flows of the vectorfields V_ζ, V_ξ respectively.

We can now define the *exponential map* $\exp_q$ at a point q in $M(p)$. This is the map from $T_q M(p)$ into $M(p)$ given by

$$\exp_q(V_\xi) = \phi^t(q, V_\xi)\Big|_{t=1}.$$

Theorem 3. *For all q in $M(p)$, the exponential map $\exp_q$ is a real analytic isomorphism between $T_q M(p) \simeq \ell_1^2$ and $M(p)$. It satisfies*

$$\kappa(\exp_q(V_\xi)) = \kappa(q) + \xi.$$

It follows that $M(p)$ is connected and simply connected.

Proof. The identity

$$\kappa(\exp_q(V_\xi)) = \kappa(q) + \xi$$

follows immediately from the definition of $\exp_q$ and the fact that V_ξ is the constant vectorfield ξ in the κ-coordinates on $M(p)$. This identity shows that the κ-coordinate system maps $M(p)$ onto ℓ_1^2, and that $\exp_q$ is the inverse of the translated map $r \to \kappa(r) - \kappa(q)$. The latter is a real analytic isomorphism between $M(p)$ and ℓ_1^2 by Theorem 1*, so the exponential map is a real analytic isomorphism between $\ell_1^2 \simeq T_q M(p)$ and $M(p)$. ∎

Corollary 1. *On each isospectral manifold $M(p)$, there is a unique even point. It is closer to the origin of L^2 than any other point on $M(p)$.*

Proof.   Consider the point

$$e = \exp_p(-V_{\kappa(p)})$$

on $M(p)$. We have

$$\kappa(e) = \kappa(p) - \kappa(p) = 0,$$

so e is even by Lemma 3.4. It is the only even point on $M(p)$, since the map μ_E is one-to-one by Theorem 3.3. It is closer to the origin in L^2 than any other point on $M(p)$, because by Lemma 2,

$$\|\exp_e(V_\xi)\| = \left(\|e\|^2 + 8 \sum_{n \geq 0} \delta_n(\cosh \xi_n - 1) \right)^{1/2}$$

$$> \|e\|$$

for all $\xi \neq 0$, and $\exp_e$ maps onto $M(p)$.   ■

Problem 5 (Another proof of Corollary 1).   Consider the differentiable function ρ on $M = M(p)$ defined by

$$\rho(q) = \|q\|^2.$$

(a)   Show that ρ attains its minimum on M.

[Hint: Pick a minimizing sequence, show that it is bounded, extract a weakly converging subsequence, and use the compactness of the functions μ_n.]

(b)   Suppose q in M is a critical point for ρ. That is, $d_q\rho \mid T_q M$ vanishes. Show that q is even.

(c)   Conclude from (a) and (b) that there is a necessarily unique even point e on M, and that it is closer to the origin of L^2 than any other point on M.

(d)   Prove directly, without using the exponential map, that M is connected.

[Hint: Minimize ρ over each component of M.]

Problem 6.   (a)   Show that an even function belongs to the Sobolev space H^1, if $M(e)$ contains an odd function.

[Hint: Use Problems 2.3 and 3.6.]

(b)   Conclude that the generic isospectral set does *not* contain an odd function.

Let us summarize the results obtained so far in the following simple picture. Every isospectral set intersects the space E of all even functions in a unique point e. Hence,

$$L^2[0, 1] = \bigcup_{e \in E} M(e),$$

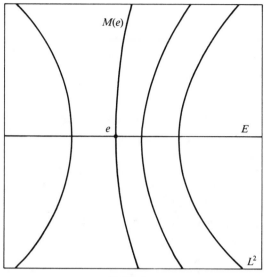

Figure 2.

and we may think of isospectral sets as fibers of a fibration of L^2 over the base E. On each fiber, we have a global coordinate system κ, which maps it onto ℓ_1^2. Its inverse is the exponential map $\exp_e$.

Problem 7. Let e be the even point on $M = M(p)$.

(a) Show that the intersection of M with the sphere of radius r centered at the origin of L^2 is analytically isomorphic to

$$
S_r = \left\{ \xi \in \ell_1^2 : 8 \sum_{n \geq 1} \delta_n(e)(\cosh \xi_n - 1) = r^2 - \|e\|^2 \right\}.
$$

(b) Show that S_r is empty if $r < \|e\|$, contains just the point 0 if $r = \|e\|$, and is the boundary of a bounded convex body in ℓ_1^2 if $r > \|e\|$.

(c) Conclude from (a) and (b) that the intersection of M with any sphere centered at the origin is connected and simply connected.

Problem 8. Let e be the even point on $M = M(p)$. Define the sum of two points q and r in M by

$$
q \oplus r = \exp_e(V_{x(q) + x(r)}).
$$

Show that e is the identity element, and that every point q has an inverse $\ominus q$.
Show that

$$\ominus q = q^*,$$

the reflection of q across the even subspace.

Another way to see how $M(p)$ lies inside L^2 is to study its projections onto
linear subspaces. A natural choice is $L^2[0, \frac{1}{2}]$ or $L^2[\frac{1}{2}, 1]$. For, the even point
in $M(p)$ is uniquely determined by its projection onto either one of these
subspaces. One might therefore expect that these projections are well
behaved at least in a small neighborhood of this even point.

Also, the restriction of a function q to the left half interval, say, determines
"half" of q. Loosely speaking, the Dirichlet spectrum determines a different
"half" of q. It may be hoped that they together determine q uniquely.

So decompose L^2 as the direct sum

$$L^2 = L^2[0, \tfrac{1}{2}] \oplus L^2[\tfrac{1}{2}, 1]$$

of square integrable functions of the left and right half intervals, and
consider the natural projections

$$\begin{array}{cc}
L^2[0, 1] & L^2[0, 1] \\
\pi_L \downarrow & \pi_R \downarrow \\
L^2[0, \tfrac{1}{2}] & L^2[\tfrac{1}{2}, 1].
\end{array}$$

Theorem 4. *For all p in L^2, the projection π_L is a real analytic isomorphism
between $M(p)$ and an open subset of $L^2[0, \frac{1}{2}]$.*

Of course, an analogous result holds for the projection π_R.

It follows that every function on the left half interval has at most one
extension to a function on the whole interval that lies in $M(p)$. This fact was
discovered by Hochstadt and Liebermann [HL].

The main ingredient of the proof of Theorem 4 is

Lemma 3.   (a)   *At every point in $M(p)$, the sequence of vectors*

$$\pi_L V_n, \qquad n \geq 1$$

is a basis of $L^2[0, \frac{1}{2}]$.

(b)   *Let q and r be any two points in M(p). Then the sequence of vectors*

$$1, \pi_L(g_n(q)g_n(r) - 1), \qquad n \geq 1$$

is a basis for $L^2[0, \frac{1}{2}]$.

(c)   *Analogous statements hold, when π_L is replaced by π_R in (a) and (b).*

Proof of Lemma 3.   (a)   Fix a point q in $M(p)$, and consider the vectors $V_n = V_n(q)$. We have

$$\frac{1}{2\pi n} V_n = 2 \sin 2\pi n x + O\left(\frac{1}{n}\right)$$

by Corollary 2.1. The sequence of trigonometric functions $2 \sin 2\pi n x$, $n \geq 1$, is an orthonormal basis for $L^2[0, \frac{1}{2}]$, and the error terms are square summable. Therefore, the statement follows from Theorem D.3, provided the vectors $\pi_L V_n$ are either linearly independent over $[0, \frac{1}{2}]$ or span $L^2[0, \frac{1}{2}]$.

Each time we have applied Theorem D.3 it was possible to verify independence directly. Here, however, this is difficult to do. To prove the linear independence of a set of functions we always needed to know their boundary values in order to construct a biorthogonal set. See for example the proofs of Theorem 2.8 and 2.9. But in the present situation, nothing is known about the values of $\pi_L V_n$ at $\frac{1}{2}$. Therefore, we are going to show that the sequence of vectors $\pi_L V_n$, $n \geq 1$, spans $L^2[0, \frac{1}{2}]$.

Suppose there is a vector u in $L^2[0, \frac{1}{2}]$, which is perpendicular to all the vectors $\pi_L V_n$. Then the entire function

$$f(\lambda) = \left\langle u, \frac{d}{dx} \pi_L y_2^2(x, \lambda) \right\rangle$$

$$= \int_0^{1/2} u(x) \frac{d}{dx} y_2^2(x, \lambda) \, dx$$

has a root at each Dirichlet eigenvalue μ_n, $n \geq 1$, of q. We show that f vanishes identically, and subsequently that u vanishes identically.

The roots of $y_2(1, \lambda)$ coincide with the Dirichlet spectrum of q and are simple. Therefore, the quotient $f(\lambda)/y_2(1, \lambda)$ is entire. By Theorem 1.3,

$$\frac{d}{dx} y_2^2(x, \lambda) = 2y_2(x, \lambda)y_2'(x, \lambda)$$

$$= \frac{\sin 2\sqrt{\lambda}\, x}{\sqrt{\lambda}} + O\left(\frac{e^{2|\text{Im}\, \sqrt{\lambda}|x}}{|\lambda|}\right)$$

for all λ, so that, by Problem 1.3,

$$f(\lambda) = \frac{1}{\sqrt{\lambda}} \int_0^{1/2} u(x) \sin 2\sqrt{\lambda}\, x\, dx + O\left(\frac{e^{|\text{Im}\,\sqrt{\lambda}|}}{|\lambda|}\right)$$

$$= o\left(\frac{e^{|\text{Im}\,\sqrt{\lambda}|}}{|\sqrt{\lambda}|}\right) + O\left(\frac{e^{|\text{Im}\,\sqrt{\lambda}|}}{|\lambda|}\right)$$

$$= o\left(\frac{e^{|\text{Im}\,\sqrt{\lambda}|}}{|\sqrt{\lambda}|}\right)$$

for all λ. Here it is important that we integrate only up to $\frac{1}{2}$. On the other hand, referring to the proof of Theorem 3.3,

$$\frac{1}{y_2(1, \lambda)} = O\left(\frac{|\sqrt{\lambda}|}{e^{|\text{Im}\,\sqrt{\lambda}|}}\right)$$

on all sufficiently large circles $|\lambda| = (n + \frac{1}{2})^2 \pi^2$. Thus,

$$\frac{f(\lambda)}{y_2(1, \lambda)} = o(1)$$

on the same circles. It follows from the maximum principle that this quotient and thus the function $f(\lambda)$ vanishes identically.

Now let $\bar{q}$ be the *even* extension of $\pi_L q$ to $[0, 1]$, and let $\bar{\mu}_n$, $n \geq 1$, be its Dirichlet eigenvalues. Since

$$y_2(x, \lambda, q) = y_2(x, \lambda, \bar{q}), \qquad 0 \leq x \leq \tfrac{1}{2},$$

and since f vanishes in particular at $\bar{\mu}_n$, we have

$$u \perp \pi_L \bar{V}_n, \qquad n \geq 1$$

for $\bar{V}_n = V_n(\bar{q})$. On $[0, 1]$, the V_n are all *odd* by Theorem 2.6. Hence, if we extend u to an *odd* function $\tilde{u}$ on $[0, 1]$, then also

$$\tilde{u} \perp \bar{V}_n, \qquad n \geq 1.$$

However, since $\bar{q}$ is even, the $\bar{V}_n$ are a basis of the space of all odd functions on $[0, 1]$ by Theorem 2.9. Consequently, $\tilde{u}$ and so u must be zero.

This shows that the vectors $\pi_L V_n$, $n \geq 1$, span $L^2[0, \frac{1}{2}]$, and the proof of (a) is complete.

(b) is proven analogously. Here one needs to show that the vectors

$$1, \pi_L(g_n(q)g_n(r) - 1), \qquad n \geq 1$$

span $L^2[0, \frac{1}{2}]$. Suppose they do not. Then there exists a function u in $L^2[0, \frac{1}{2}]$ which is perpendicular to all of them. It follows as in the proof of (a) that the function

$$f(\lambda) = \int_0^{1/2} u(x)y_2(x, \lambda, q)y_2(x, \lambda, r)\, dx$$

vanishes not only at the eigenvalues of q and r, but vanishes identically.[3]

Now extend $\pi_L q$ to a function w on $[0, 1]$ by setting

$$w = q \text{ on } [0, \tfrac{1}{2}], \qquad w = r^* \text{ on } [\tfrac{1}{2}, 1].$$

Then w^* is an extension of $\pi_L r$ to $[0, 1]$, and so, for $0 \le x \le \frac{1}{2}$,

$$y_2(x, \lambda, q) = y_2(x, \lambda, w), \quad y_2(x, \lambda, r) = y_2(x, \lambda, w^*).$$

Since f vanishes in particular at $\mu_n(w) = \mu_n(w^*)$, $n \ge 1$, it follows that u is perpendicular to all the vectors

$$1, \pi_L(g_n(w)g_n(w^*) - 1), \qquad n \ge 1.$$

The vectors $g_n(w)g_n(w^*)$ are all even on $[0, 1]$. Hence the even extension $\bar{u}$ of u is perpendicular on $[0, 1]$ to all the vectors

$$1, g_n(w)g_n(w^*) - 1, \qquad n \ge 1.$$

But these vectors are a basis for the space E of all even functions by the next Problem. Consequently, $\bar{u}$ and so u must be zero.

This shows that the vectors $1, \pi_L(g_n(q)g_n(r) - 1)$ span $L^2[0, \frac{1}{2}]$.

(c) The analogous statements for the projection on the right half interval follow from (a) and (b) by reflection. ∎

Problem 9.   Show that at every point q in L^2, the vectors $1, g_n(q)g_n(q^*) - 1$, $n \ge 1$, are a basis for E.

[Hint: These vectors are even since $g_n(q^*) = (-1)^n g_n^*(q)$. To prove their independence, calculate the inner product between these vectors and

$$\frac{d}{dx}(y_1(x, \mu_n(q), q)y_2(x, \mu_n(q), q^*)), \qquad n \ge 1$$

as in the proof of Theorem 2.8. Otherwise imitate the proof of Theorem 2.9.]

[3] The proof is actually easier than before because now

$$f(\lambda) = O\left(\frac{e^{|\text{Im }\sqrt{\lambda}|}}{|\lambda|}\right).$$

Proof of Theorem 4.   The restriction $\pi_L \mid M$ of π_L to $M = M(p)$ is clearly real analytic, and its derivative at q is the restriction of $d_q \pi_L$ to $T_q M$. It is the linear map from $T_q M$ into $L^2[0, \frac{1}{2}]$ such that

$$V_n \rightarrow \pi_L V_n, \quad n \geq 1.$$

Since the vectors $\pi_L V_n$ are a basis for $L^2[0, \frac{1}{2}]$ by Lemma 3, this derivative is boundedly invertible, and $\pi_L \mid M$ is a local real analytic isomorphism by the inverse function theorem.

We show that $\pi_L \mid M$ is one-to-one. Suppose q and r are two points on M which agree on the left half interval. Cross multiplying the differential equations for $g_n(q)$ and $g_n(r)$ and taking their difference, we get

$$0 = \langle q - r, g_n(q)g_n(r) \rangle$$

$$= \langle \pi_R(q - r), \pi_R(g_n(q)g_n(r)) \rangle$$

for all $n \geq 1$. Moreover,

$$0 = [q] - [r]$$

$$= \langle q - r, 1 \rangle$$

$$= \langle \pi_R(q - r), 1 \rangle.$$

Part (b) of Lemma 3 for the projection π_R now implies that

$$\pi_R(q - r) = 0.$$

Hence, $q = r$. ■

Problem 10.   (a)   Show that the projection of $M(p)$ into $L^2[\frac{1}{4}, \frac{1}{2}] \oplus L^2[\frac{3}{4}, 1]$ is a local real analytic isomorphism at the event point.

[Hint: Use the fact that at an even point, the functions g_n^2, $n \geq 1$, are a basis for the space E of even functions.]

(b)   What about the projection into $L^2[\frac{1}{4}, \frac{3}{4}]$?

Problem 11.   Fix u in $L^2[0, \frac{1}{2}]$ and consider the function ρ_u on L^2 defined by

$$\rho_u(q) = \int_0^{1/2} (q - u)^2 \, dx.$$

Show that u lies in the projection of $M(p)$ into $L^2[0, \frac{1}{2}]$ if and only if the restriction of ρ_u to $M(p)$ has a critical point. Show that such a critical point is necessarily unique.

The question remains whether $M(p)$ ever projects *onto* $L^2[0, \frac{1}{2}]$. The answer is, it *never* does. In other words, there always is a function in $L^2[0, \frac{1}{2}]$ which can not be extended in $L^2[0, 1]$ to a function in $M(p)$.

Such functions can be written down explicitly. Fix q in $M(p)$ and $n \geq 1$. As will be seen in Chapter 5, as $t \to \infty$, the curve $\phi^t(q, V_n) = \phi_n^t(q)$ converges in $L^2[0, \frac{1}{2}]$ to the function

$$\phi_n^\infty(q) = q - 2 \frac{d^2}{dx^2} \log \int_x^1 g_n^2(s, q) \, ds,$$

but tends to infinity in $L^2[\frac{1}{2}, 1]$. It follows from Theorem 4 that $\phi_n^\infty(q)$ is not contained in the range of the projection of $M(p)$ into $L^2[0, \frac{1}{2}]$, but lies on its boundary. For otherwise, $\phi_n^\infty(q)$ would be an interior point. In this case, applying the inverse of the isomorphism $\pi_L \mid M(p)$,

$$\phi_n^t(q) = \pi_L^{-1}(\pi_L(\phi_n^t(q)))$$

must be bounded in $L^2[0, 1]$ for $t \to \infty$ by continuity. This is a contradiction.

5   Explicit Solutions

In the last century an ordinary differential equation was "solved" when all its solutions could be expressed as finite or perhaps infinite combinations of known functions. For example, as a power series whose coefficients satisfy a manageable recurrence relation. The work of Poincaré and others, however, led to the realization that this is generally impossible. As a consequence, qualitative and geometric methods were developed, and the search for explicit solutions was all but given up.

In 1965 it therefore came as a surprise that certain interesting nonlinear evolution equations can be solved in closed form. One well known example is the Korteweg-de Vries equation describing the propagation of water waves in a canal. This discovery renewed interest in explicit solutions, and quite a number of such "integrable systems" were subsequently found.

We shall see that the differential equation

$$\frac{d}{dt} q = V_{\xi}(x, q)$$

on L^2 is also "integrable". Exploiting an observation that goes back to Darboux [Da1], we shall derive a closed formula for its solution curves. For the vectorfields $V_n = 2(d/dx)g_n^2$ this formula reads

$$\phi'(q, V_n) = q - 2\frac{d^2}{dx^2}\log\theta_n(x, t, q),$$

where

$$\theta_n(x, t, q) = 1 + (e^t - 1)\int_x^1 g_n^2(s, q)\, ds.$$

In addition, there is a closed formula for the exponential map $\exp_q$ between $\ell_1^2 \simeq T_q M(p)$ and $M(p)$. Namely,

$$\exp_q(V_\xi) = q - 2\frac{d^2}{dx^2}\log\det\Theta(x, \xi, q),$$

where Θ is an infinite matrix with entries

$$\theta_{ij}(x, \xi, q) = \delta_{ij} + (e^{\xi_i} - 1)\int_x^1 g_i(s, q)g_j(s, q)\, ds.$$

The infinite determinant is defined later on as a limit of finite dimensional determinants.

There are two steps in the derivation of these formulae. First, the solution curves of the vectorfields V_n are determined. Then they are combined. Along the way, a general method is developed that will be used again in the next chapter.

The heart of the matter is the following algebraic fact.

Lemma 1. *Pick a real number μ, and let g be a nontrivial solution of*

(1) $-y'' + qy = \lambda y$

for $\lambda = \mu$. If f is a nontrivial solution of (1) for $\lambda \neq \mu$, then

$$\frac{1}{g}[g, f]$$

is a nontrivial solution of

(2) $-y'' + \left(q - 2\frac{d^2}{dx^2}\log g\right)y = \lambda y$

for the same λ. Also, for $\lambda = \mu$, the general solution of (2) is given by

$$\frac{1}{g}\left(a + b\int_0^x g^2(s)\, ds\right),$$

where a and b are arbitrary constants. In particular, $1/g$ is a solution.

If g has roots in $[0, 1]$, then equation (2) is understood to hold between them. Notice that the sign of g is unimportant for the logarithmic derivative $(d/dx) \log g = g'/g$.

Proof. Lemma 1 can be established by direct calculation. However, there is a more systematic proof based on the observation that the second order operator $-(d^2/dx^2) + q - \mu$ can be written as the product of two first order operators. This observation, sometimes attributed to d'Alembert, is traditionally referred to as the "method of reduction".

Let $A = g(d/dx)(1/g)$ and $A^* = -(1/g)(d/dx)g$. The operator A^* is the formal adjoint of A, if g and μ are real. Using the differential equation for g we find[1]

$$A^*A = -\frac{d^2}{dx^2} + q - \mu.$$

It follows that equation (1) can be rewritten as

(1')
$$A^*Ay = (\lambda - \mu)y.$$

Similarly,

$$AA^* = -\frac{d^2}{dx^2} - \frac{g''}{g} + 2\left(\frac{g'}{g}\right)^2$$

$$= -\frac{d^2}{dx^2} + \frac{g''}{g} + 2\left(\left(\frac{g'}{g}\right)^2 - \frac{g''}{g}\right)$$

$$= -\frac{d^2}{dx^2} + \left(q - 2\frac{d^2}{dx^2}\log g\right) - \mu,$$

so that equation (2) becomes

(2')
$$AA^*y = (\lambda - \mu)y.$$

The lemma is now easy to prove. Applying A to both sides of equation (1') we obtain

$$AA^*Ay = (\lambda - \mu)Ay.$$

So, if y is a solution of (1'), then Ay is a solution of (2'). Consequently, if f is a solution of (1) for $\lambda \neq \mu$, then

$$Af = g\frac{d}{dx}\frac{f}{g} = f' - f\frac{g'}{g} = \frac{1}{g}[g, f]$$

[1] For some insight into this identity see the next Problem.

is a solution of (2). Moreover, Af is a nontrivial solution when f is non-trivial. Otherwise, the Wronskian $[g, f]$ vanishes identically, which implies

$$0 = [g, f]' = (\mu - \lambda)gf.$$

But then $\lambda = \mu$, since g and f are nontrivial.

To prove the second statement, suppose h is a solution of equation (2) for $\lambda = \mu$. Then

$$AA^*h = -g \frac{d}{dx} \frac{1}{g^2} \frac{d}{dx} gh = 0,$$

or

$$\frac{d}{dx} gh = bg^2$$

for some constant b. The result follows from another integration. ■

Problem 1. (a) Show that the most general real, formally self adjoint second order ordinary differential operator is of the form

$$\frac{d}{dx} p(x) \frac{d}{dx} + r(x).$$

(b) Deduce from (a), by calculating the leading coefficients, that A^*A and AA^* are both of the form $-(d^2/dx^2) + r(x)$. Here, as in the proof of Lemma 1, $A = g(d/dx)(1/g)$, where g is a nontrivial solution of (1), and $A^* = -(1/g)(d/dx)g$, the formal adjoint of A for real g.

(c) Derive the identities

$$A^*A = -\frac{d^2}{dx^2} + \frac{g''}{g}, \qquad AA^* = -\frac{d^2}{dx^2} + g\left(\frac{1}{g}\right)''$$

from $Ag = 0$ and $A^*(1/g) = 0$ respectively. Use the differential equation for g to show that (1) and (2) are equivalent to (1') and (2') respectively.

Lemma 1 was known to Gaston Darboux. A proof appeared in his 1882 Comptes rendus note "Sur une proposition rélative aux équations linéaires". In article 408 of his "Theory of Surfaces" [Da2] Darboux calls it a "curious theorem of analysis" and writes:

"This proposition, which is easy to verify directly, evidently permits one to associate to every equation of the form (1), that one knows how to integrate for all values of λ, an infinite sequence of differential equations of the same form that one also knows how to integrate for all values of the parameter λ. Each passage from one equation to the next introduces two new arbitrary constants[2]; in general, the successive equations differ more and more from the initial form and become more and more complicated. There are, however, exceptional cases in which the form of the equation is preserved, when one chooses the particular solutions appropriately.''

Actually, Darboux did not express the new differential equation in terms of the second logarithmic derivative of g. Instead he wrote the equation in the form

$$-y'' + g\left(\frac{1}{g}\right)'' y = (\mu - \lambda)y.$$

It is the addition law of the logarithm that makes iteration of Lemma 1 practical.

Lemma 2. *Pick real numbers μ and v. Let g be a nontrivial solution of (1) for $\lambda = \mu$ and h a nontrivial solution of (2) for $\lambda = v$. If f is a nontrivial solution of (1) for $\lambda \neq \mu, v$, then*

$$\frac{1}{h}\left[h, \frac{1}{g}[g, f]\right] = (\mu - \lambda)f - \frac{1}{g}[g, f]\frac{d}{dx}\log(gh)$$

is a nontrivial solution of

(3) $$-y'' + \left(q - 2\frac{d^2}{dx^2}\log(gh)\right)y = \lambda y$$

for the same λ. Also, $1/h$ is a nontrivial solution of (3) for $\lambda = v$.

Equation (3) is understood to hold between the roots of gh.

Proof. Apply Lemma 1 to equation (2) with h in place of g and $[g_1 f]/g$ in place of f. ∎

We are ready to determine the solution curves of the vectorfield V_n.

Theorem 1. *The solution curve of the vectorfield $V_n = 2(d/dx)g_n^2$ with initial value q on $M(p)$ is given by*

$$\phi^t(q, V_n) = q - 2\frac{d^2}{dx^2}\log\theta_n(x, t, q), \qquad -\infty < t < \infty,$$

[2] The initial values of g [the authors].

where

$$\theta_n(x, t, q) = 1 + (e^t - 1) \int_x^1 g_n^2(s) \, ds.$$

Moreover,

$$g_j(x, \phi^t) = e^{\delta_{jn} t/2} \left(g_j - (e^t - 1) \frac{g_n}{\theta_n} \int_x^1 g_j(s) g_n(s) \, ds \right),$$

where $\phi^t = \phi^t(q, V_n)$ and $g_j = g_j(x, q)$ for $j \geq 1$.

Observe that

$$g_n(x, \phi^t) = e^{t/2} \left(g_n - \frac{g_n}{\theta_n} (\theta_n - 1) \right)$$

$$= e^{t/2} \frac{g_n}{\theta_n}.$$

Proof.   Fix q in $M(p)$ and a positive integer n. By Lemma 1,

$$h = \frac{1}{g_n} \left(1 + c \int_0^x g_n^2(s) \, ds \right)$$

is a solution of equation (2) for $\lambda = \mu_n(q)$ and $g = g_n(x, q)$. Here, c is a real parameter. The idea of the proof is to apply Lemma 2 to this solution h of equation (2) and the solutions

$$g = g_n, \qquad f = g_j, \qquad j \neq n$$

of equation (1).
Set

$$\theta_{n,c} = gh = g_n h$$

$$= 1 + c \int_0^x g_n^2(s) \, ds.$$

For $c > -1$, the function $\theta_{n,c}$ is strictly positive on $[0, 1]$. For,

$$0 \leq \int_0^x g_n^2(s) \, ds \leq 1, \qquad 0 \leq x \leq 1$$

and so

$$1 + c \int_0^x g_n^2(s) \, ds \geq \begin{cases} 1, & c \geq 0 \\ 1 + c, & 0 > c > -1 \end{cases} > 0.$$

Set

$$q_c = q - 2\frac{d^2}{dx^2}\log\theta_{n,c}.$$

For $c > -1$, the function q_c belongs to L^2, since $\theta_{n,c}$ is strictly positive. Furthermore, for $j \geq 1$, let[3]

$$g_{j,c} = g_j - \frac{1}{g_n}\frac{[g_n, g_j]}{\mu_n - \mu_j}\frac{d}{dx}\log\theta_{n,c}$$

$$= g_j - c\frac{g_n}{\theta_{n,c}}\int_0^x g_j g_n \, ds.$$

Once again, we dropped the arguments x and q. In the second line the identity $[g_n, g_j] = (\mu_n - \mu_j)\int_0^x g_n g_j \, ds$ is used. Notice that

$$g_{n,c} = \frac{1}{h} = \frac{g_n}{\theta_{n,c}}.$$

By Lemma 2, the function $g_{j,c}$, $j \geq 1$, is a genuine nontrivial solution of

$$-y'' + q_c y = \mu_j(q)y.$$

Moreover, it vanishes at 0 and 1. Therefore, $\mu_j(q)$ is a Dirichlet eigenvalue for q_c. In fact, $g_{j,c}$ has the same number of roots in $[0, 1]$ as $g_{j,0} = g_j$ by Lemma 2.3. It follows that $g_{j,c}$ is a constant multiple of the jth normalized eigenfunction of q_c, and consequently that

$$\mu_j(q_c) = \mu_j(q)$$

for all $j \geq 1$. That is, q_c has the same Dirichlet spectrum as q for all $c > -1$.

We determine the normalized eigenfunctions of q_c. Squaring the second expression for $g_{j,c}$ and using $\theta'_{n,c} = cg_n^2$, one finds after a short calculation that

$$g_{j,c}^2 = g_j^2 - c\frac{d}{dx}\left(\frac{1}{\theta_{n,c}}\left(\int_0^x g_j g_n \, ds\right)^2\right).$$

Hence,

$$\int_0^1 g_{j,c}^2 \, ds = 1 - \frac{1}{1+c}\delta_{jn} = \frac{1}{1+c\delta_{jn}}.$$

[3] Keep in mind that q_c and $g_{j,c}$ also depend on the index n, which is suppressed for simplicity.

Also, $g'_{j,c}(0) = g'_j(0)$ is positive for all $j \geq 1$. Therefore,

$$g_j(q_c) = \sqrt{1 + c\delta_{jn}}\, g_{j,c}$$

for all $j \geq 1$.

So far we have constructed a path q_c, $c > -1$, that lies on $M(p)$. It satisfies

$$\kappa_j(q_c) = \log \left| \frac{g'_j(1, q_c)}{g'_j(0, q_c)} \right|$$

$$= \log \left| \frac{g'_{j,c}(1)}{g'_{j,c}(0)} \right|$$

$$= \log \left| \frac{g'_j(1)}{g'_j(0)} \frac{1}{1 + c\delta_{jn}} \right|$$

$$= \kappa_j(q) - \delta_{jn} \log(1 + c), \qquad j \geq 1.$$

On the other hand, the solution curve $\phi^t(q, V_n)$ is the unique path on $M(p)$ that satisfies

$$\kappa_j(\phi^t(q, V_n)) = \kappa_j(q) + \delta_{jn} t, \qquad j \geq 1.$$

Thus, if we choose $c = e^{-t} - 1 > -1$, $-\infty < t < \infty$, then

$$\kappa_j(q_c) = \kappa_j(q) + \delta_{jn} t, \qquad j \geq 1.$$

Moreover,

$$\theta_{n,c} = 1 + (e^{-t} - 1)\left(1 - \int_x^1 g_n^2\, ds\right)$$

$$= e^{-t}\left(1 + (e^t - 1) \int_x^1 g_n^2\, ds\right)$$

$$= e^{-t}\theta_n(x, t).$$

Therefore,

$$\phi^t(q, V_n) = q_c$$

$$= q - 2\frac{d^2}{dx^2} \log e^{-t}\theta_n(x, t)$$

$$= q - 2\frac{d^2}{dx^2} \log \theta_n(x, t).$$

The proof of the first part of Theorem 1 is complete. The second part is obtained by a straightforward calculation. ∎

Problem 2 (Proof of Theorem 1 by direct calculation).

(a) Show that $\psi_n^t(q) = q - 2(d^2/dx^2)\log\theta_n$ is a differentiable curve in L^2.

(b) Show that $\psi_n^0(q) = q$ and

$$\frac{d}{dt}\psi_n^t(q) = -2\frac{d^2}{dx^2}\frac{d}{dt}\log\theta_n = 2\frac{d}{dx}h_n^2,$$

where

$$h_n = e^{t/2}\frac{g_n}{\theta_n},$$

and g_n, θ_n are evaluated at q.

(c) Show that

$$h_n = g_n(\psi_n^t(q)).$$

To this end check that

(1) $-h_n'' + \psi_n^t(q)h_n = \mu_n(q)h_n$;
(2) h_n has exactly $n + 1$ roots in $[0, 1]$ including 0 and 1
(3) h_n has positive derivative at 0
(4) h_n has norm one in L^2.

[Hint: For the last point use the identity $h_n^2 = (e^t/e^t - 1)(1/\theta_n)'$.]

(d) Conclude that ψ_n^t is a solution curve of the vectorfield V_n with initial value q, hence $\psi_n^t = \phi^t(q, V_n)$ by uniqueness.

The proof of Theorem 1 is not difficult, but at the same time, not particularly transparent. We now try to make the whole approach seem more "natural" in hindsight by exposing an underlying "trick".

First a general remark. Consider two operators A and B on a Hilbert space. In finite dimensions, AB and BA always have the same spectrum. This is easy to see. If A is invertible, then

$$\det(AB - \lambda) = \det(A^{-1})\det(AB - \lambda)\det(A) = \det(BA - \lambda).$$

Invertible matrices are dense, so the identity $\det(AB - \lambda) = \det(BA - \lambda)$ holds in general. Thus, AB and BA have the same characteristic polynomial and hence the same spectrum.

In infinite dimensions, AB and BA have the same spectrum *away from zero*, provided they are bounded. That is, 0 may belong to the spectrum of

AB, but not to the spectrum of BA, or vice versa. This is also easy to see. One shows, by multiplying out, that for each $\lambda \neq 0$ in the resolvent set of AB, the operator

$$B(AB - \lambda)^{-1}A - I$$

is a two sided inverse of $BA - \lambda$, so that λ also belongs to the resolvent set of BA. The opposite inclusion holds by symmetry. Thus, AB and BA have the same resolvent set away from zero, hence also the same spectrum away from zero.

In a special case, there is a generalization to unbounded operators. If A is a closed operator with adjoint A^*, then A^*A and AA^* have the same spectrum away from zero [De, DeT].

These observations suggest a way of constructing families of functions with the same Dirichlet spectrum.[4]

Pick a function q in L^2 and let g_m, $m \geq 1$, be its Dirichlet eigenfunctions. Fix $n \geq 1$. As in the proof of Lemma 1, the operator $-(d^2/dx^2) + q - \mu_n$ may be factored as A^*A, where

$$A = g_n \frac{d}{dx} \frac{1}{g_n}, \qquad \text{and} \qquad A^* = -\frac{1}{g_n} \frac{d}{dx} g_n$$

is its formal adjoint. Interchanging the factors, one may expect that

$$AA^* = -\frac{d^2}{dx^2} + \left(q - 2 \frac{d^2}{dx^2} \log g_n \right) - \mu_n$$

has the same "Dirichlet spectrum" as A^*A with the possible exception of 0.

Formally, this is so. The functions Ag_m, $m \neq n$, are "Dirichlet eigenfunctions" of AA^*: they are nontrivial and vanish at 0 and 1 by l'Hospital's rule. On the other hand, by Lemma 1 the general solution of $AA^*f = 0$ is unbounded at 0 or 1 unless it vanishes identically. So, 0 is not a "Dirichlet eigenvalue". Unfortunately, the coefficient in AA^* does not belong to L^2, and the "eigenfunctions" Ag_m are not defined at the roots of g_n.

To overcome these difficulties and construct functions with the same spectrum as q, let us factor once more. In the hope of restoring 0 as an eigenvalue, choose a solution

$$h = \frac{1}{g_n} \left(a + b \int_0^x g_n^2(s)\, ds \right)$$

<hr>

[4] We thank P. Deift for helpful discussions on this point.

of $AA^*f = 0$, and write

$$AA^* = B^*B,$$

where $B = h(d/dx)(1/h)$ and $B^* = -(1/h)(d/dx)h$. Once again, permuting the factors, the "Dirichlet spectrum" of

$$BB^* = -\frac{d^2}{dx^2} + \left(q - 2\frac{d^2}{dx^2}\log g_n h\right) - \mu_n$$

ought to be the same away from zero.

Indeed, for appropriate choices of a and b, everything turns out alright, as we have seen in the proof of Theorem 1. In particular,

$$\frac{1}{h} = \frac{g_n}{a + b\int_0^x g_n^2\,ds}$$

is a genuine solution of $BB^*f = 0$, when the numerator is positive, so that 0 is in fact a Dirichlet eigenvalue.

The process of factoring and permuting the factors is familiar. Consider for example the QR-algorithm well known to numerical analysts. Here, a non-singular, real symmetric matrix A is factored as

$$A = QR,$$

where Q is an orthogonal matrix, and $R = Q^{-1}A$ is an upper triangular matrix. Q is the matrix, whose columns are obtained by orthogonalizing the columns of A from left to right. Interchanging Q and R, one obtains a new matrix

$$A_1 = RQ = Q^{-1}AQ,$$

which is similar to A. Repeating this process inductively, one writes

$$A_n = Q_n R_n,$$

$$A_{n+1} = R_n Q_n = Q_n^{-1}A_n Q_n.$$

One can show that the matrices A_n converge to a diagonal matrix D and the products $QQ_1 \cdots Q_n$ to an orthogonal matrix U such that

$$U^{-1}AU = D.$$

This algorithm provides a numerically efficient way of determining the spectrum of a symmetric matrix.

The explicit formulas of Theorem 1 lend themselves to numerical calculations. They are particularly simple for the initial value $q = 0$, where the Dirichlet eigenfunctions are just trigonometric functions:

$$\phi^t(0, V_n) = -2\frac{d^2}{dx^2}\log\left(1 + (e^t - 1)\int_x^1 2\sin^2\pi ns\,ds\right).$$

Sample programs in FORTRAN for this case are given in Appendix G. They were used to produce the figures in this and the next chapter.

We plot the points on the solution curves of V_1 and V_2 with initial value 0 for

(*) $t = k \cdot 0.5, \qquad k = 0, \ldots, 5.$

That is, we increment the first and second κ-coordinate respectively in steps of 0.5 starting at $q = 0$.

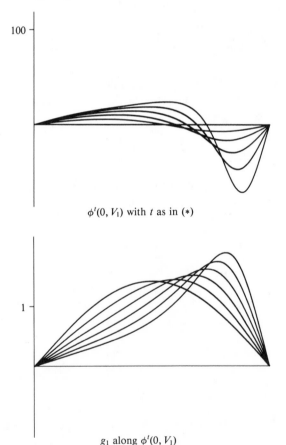

$\phi^t(0, V_1)$ with t as in (*)

g_1 along $\phi^t(0, V_1)$

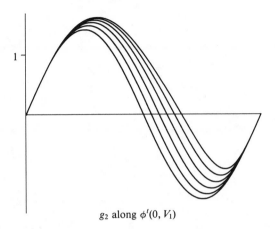

g_2 along $\phi^t(0, V_1)$

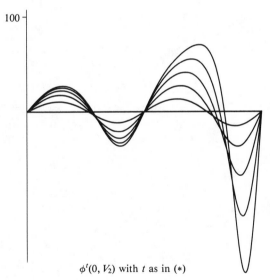

$\phi^t(0, V_2)$ with t as in $(*)$

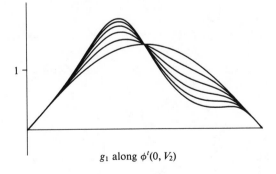

g_1 along $\phi^t(0, V_2)$

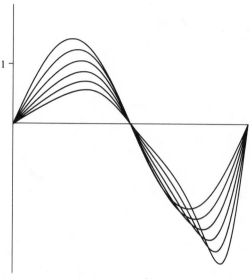

$$g_2 \text{ along } \phi'(0, V_2)$$

It is a consequence of Lemma 4.2 that the functions $\phi'(q, V_n)$ become unbounded as t becomes unbounded. A quick look at the plots suggests that this happens in a very specific way. As $t \to \infty$, these functions seem to "explode" at the right hand endpoint of the interval $[0, 1]$, while they appear to converge to a finite limit on every interval $[0, 1 - \varepsilon]$ strictly contained in $[0, 1]$.

This is the case. Fix q in L^2, and let g_n be its nth eigenfunction, $n \geq 1$. The formula of Theorem 1 can be rewritten as

$$\phi'(q, V_n) = q - 2 \frac{d^2}{dx^2} \log\left(1 + (e^t - 1) \int_x^1 g_n^2 \, ds \right)$$

$$= q - 2 \frac{d^2}{dx^2} \log\left(\frac{1}{e^t - 1} + \int_x^1 g_n^2 \, ds \right).$$

The integral $\int_x^1 g_n^2 \, ds$ is bounded away from zero as long as x is bounded away from 1, whereas at 1 it tends to zero like a power of $1 - x$. Consequently, as $t \to \infty$, the curve $\phi'(q, V_n)$ converges to the function

$$q_n^+ = q - 2 \frac{d^2}{dx^2} \log \int_x^1 g_n^2 \, ds$$

in $L^2[0, 1 - \varepsilon]$ for every $0 < \varepsilon < 1$, but tends to infinity in $L^2[1 - \varepsilon, 1]$ for every $0 < \varepsilon < 1$.

By the way, this completes the argument given at the end of Chapter 4 showing that the projection of any isospectral set into $L^2[0, \frac{1}{2}]$ is never onto. Similarly, as $t \to -\infty$, the same curve converges to the function

$$q_{\bar{n}} = q - 2\frac{d^2}{dx^2}\log\int_0^x g_n^2\, ds$$

in $L^2[\varepsilon, 1]$ but diverges in $L^2[0, \varepsilon]$ for every $0 < \varepsilon < 1$.

Problem 3.   Let q be in L^2 and let g_n be its nth eigenfunction, $n \geq 1$. Set $\phi_n^t(q) = \phi^t(q, V_n)$.
   (a)   Show that the roots of $g_n(\phi_n^t(q))$ are independent of t.
   (b)   Fix two roots $0 \leq a < b \leq 1$ of g_n. Verify the identity

$$\frac{d}{dt}\|\phi_n^t(q)\|_{L^2_{[a,b]}}^2 = 4\frac{e^t(g_n'(x, q))^2}{\theta_n^2(x, t, q)}\Big|_a^b.$$

[Hint: Proceed as in the proof of Lemma 4.1 and use the identity following Theorem 1.]
   (c)   Check that

$$\int_0^t \frac{e^s\, ds}{\theta_n^2(x, s)} = \frac{e^t - 1}{\theta_n(x, t)},$$

and conclude that

$$\|\phi_n^t(q)\|_{L^2_{[a,b]}}^2 = \|q\|_{L^2_{[a,b]}}^2 + 4\left(\frac{e^t - 1}{\theta_n(x, t, q)}(g_n'(x))^2\right)\Big|_a^b.$$

   (d)   Suppose that $b < 1$. By the remark above, the curve $\phi_n^t(q)$ has a limit q_n^+ in $L^2[a, b]$ as $t \to \infty$. Show that

$$\|q_n^+\|_{L^2_{[a,b]}}^2 = \|q\|_{L^2_{[a,b]}}^2 + 4\frac{(g_n'(x))^2}{\int_x^1 g_n^2\, ds}\Big|_a^b.$$

In particular, if e is even and n is even, show that

$$\|e_n^+\|_{L^2_{[a,b]}}^2 = \|e\|_{L^2_{[a,b]}}^2 + 8(g_n'(\tfrac{1}{2}))^2 - 4(g_n'(0))^2.$$

Combining the explicit solutions for the vectorfields V_n, it is now possible to obtain the formula for $\exp_q(V_\xi)$ stated at the beginning of this chapter.
   For $\xi \in \ell_1^2$, consider the infinite matrix

$$\Theta(x, \xi, q) = (\theta_{ij})$$

where

$$\theta_{ij}(x, \xi, q) = \delta_{ij} + (e^{\xi_i} - 1) \int_x^1 g_i(s, q)g_j(s, q) \, ds.$$

We define the determinant of Θ as the limit of the determinants of its principal minors. That is,

$$\det \Theta = \lim_{n \to \infty} \det \Theta^{(n)},$$

where

$$\Theta^{(n)} = (\theta_{ij})_{1 \le i, j \le n}.$$

This simple definition of det Θ agrees with the invariantly defined Fredholm determinant of Θ. For the argument see the next Problem.

Theorem 2. *For q in L^2 and ξ in ℓ_1^2,*

$$\exp_q(V_\xi) = q - 2\frac{d^2}{dx^2} \log \det \Theta(x, \xi, q).$$

It is implicit in the statement of the theorem that the determinant of Θ always exists and never vanishes.

If p is given, then, at least in principle, one can obtain all the eigenfunctions of p and "write down" the determinant of Θ at p. Hence, Theorem 2 provides an explicit description of the set

$$M(p) = \{\exp_p(V_\xi): \xi \in \ell_1^2\}$$

of all functions with the same Dirichlet spectrum as p.

Proof. Let

$$q_n = \exp_q(V_{\tau_n \xi}), \qquad n \ge 1,$$

where $\tau_n \xi = (\xi_1, \ldots, \xi_n, 0, \ldots)$. The truncations $\tau_n \xi$ converge to ξ in ℓ_1^2, so the vectors $V_{\tau_n \xi}$ converge to V_ξ in L^2. It follows that

$$q_n \to \exp_q(V_\xi)$$

strongly in L^2 by the continuity of the exponential map.

On the other hand, we have

$$q_n = \exp_{q_{n-1}}(\xi_n V_n)$$

$$= \phi^{\xi_n}(q_{n-1}, V_n), \qquad n \ge 1,$$

with $q_0 = q$, since q_{n-1} and q_n differ only in their nth κ-coordinate by ξ_n. Applying Theorem 1 to $q_0, \ldots, q_{n-1}$,

$$q_n = q_0 - \sum_{k=1}^{n} 2 \frac{d^2}{dx^2} \log \theta_k(x, \xi_k, q_{k-1})$$

$$= q_0 - 2 \frac{d^2}{dx^2} \log \prod_{k=1}^{n} \theta_k(x, \xi_k, q_{k-1}).$$

In Appendix F, the product on the right is identified with the determinant of the nth minor of Θ by Gaussian elimination. That is,

$$\prod_{k=1}^{n} \theta_k(x, \xi_k, q_{k-1}) = \det \Theta^{(n)}(x, \xi, q).$$

Hence,

$$q_n = q_0 - 2 \frac{d^2}{dx^2} \log \det \Theta^{(n)}(x, \xi, q).$$

Also, at $x = 0$, $\det \Theta^{(n)}$ has value $\exp(\sum_{k=1}^{n} \xi_k)$ and a vanishing first derivative. So integrating twice with respect to x and exponentiating we obtain

$$\det \Theta^{(n)} = \exp\left(\sum_{k=1}^{n} \xi_k - \tfrac{1}{2} \int_{0}^{x} (x - t)(q_n - q_0)(t)\, dt \right).$$

The convergence of the q_n and the absolute convergence of $\sum_{k>1} \xi_k$ imply that $\det \Theta^{(n)}$ converges uniformly on $[0, 1]$ to a positive function. Consequently, $\det \Theta$ always exists, never vanishes, and the formula for $\exp_q(V_\xi)$ holds as stated. ■

Corollary 1. *For q in L^2 and ξ in ℓ_1^2,*

$$\phi^t(q, V_\xi) = q - 2 \frac{d^2}{dx^2} \log \det \Theta(x, t\xi, q).$$

For $q = 0$, the formula of Theorem 2 becomes

$$\exp_0(V_\xi) = -2 \frac{d^2}{dx^2} \log \Theta(x, \xi, 0),$$

where

$$\Theta(x, \xi, 0) = I + \left((e^{\xi_i} - 1) \int_x^1 2 \sin(\pi i s) \sin(\pi j s) \, ds \right).$$

This formula is used to plot some points on $M(0)$.

In the figure captions only the nonzero ξ-coordinates are displayed.

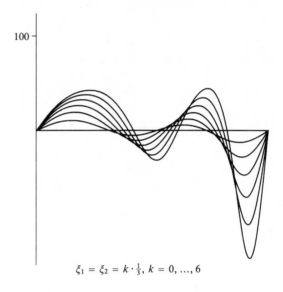

$$\xi_1 = \xi_2 = k \cdot \tfrac{1}{3}, \, k = 0, \ldots, 6$$

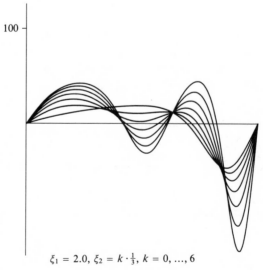

$$\xi_1 = 2.0, \, \xi_2 = k \cdot \tfrac{1}{3}, \, k = 0, \ldots, 6$$

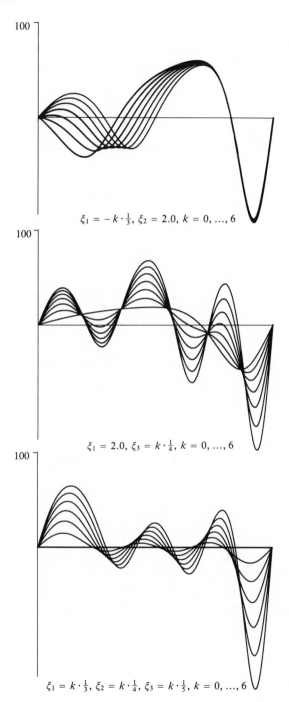

$\xi_1 = -k \cdot \frac{1}{3},\ \xi_2 = 2.0,\ k = 0, ..., 6$

$\xi_1 = 2.0,\ \xi_3 = k \cdot \frac{1}{4},\ k = 0, ..., 6$

$\xi_1 = k \cdot \frac{1}{3},\ \xi_2 = k \cdot \frac{1}{4},\ \xi_3 = k \cdot \frac{1}{5},\ k = 0, ..., 6$

Problem 4 (The Fredholm determinant of Θ).   Let A be a linear operator on a Hilbert space. If A is trace class, then the Fredholm determinant of $I + A$ is defined by the identity

$$\det(I + A) = \sum_{k \geq 0} \text{tr}(\Lambda^k(A)).$$

where $\Lambda^k(A)$ is the kth exterior power of A. See [RS] for the details as well as for the definition of the trace class norm $\|\cdot\|_1$ and the Hilbert Schmidt norm $\|\cdot\|_2$.

We will use the following facts proven in [RS]. The determinant of $I + A$ is a continuous function of A with respect to the trace class norm. If B, C are both Hilbert Schmidt, then BC is trace class, and $\|BC\|_1 \leq \|B\|_2 \|C\|_2$.

(a)   Write $\Theta - I$ as the product of the two matrices

$$B = (\sqrt{|\xi_i|}\, \delta_{ij})$$

and

$$C = \left(\frac{e^{\xi_i} - 1}{\sqrt{|\xi_i|}} \int_x^1 g_i(s) g_j(s)\, ds \right).^5$$

Show that both matrices are Hilbert Schmidt on ℓ^2. It follows that $\Theta - I$ is trace class on ℓ^2, hence the determinant of Θ according to the above definition exists.

[Hint: Use that

$$\sum_{j \geq 1} \left(\int_x^1 g_i(s) g_j(s)\, ds \right)^2 = \sum_{j \geq 1} \langle \mathbb{1}_{[x,1]} g_i, g_j \rangle^2$$

$$= \| \mathbb{1}_{[x,1]} g_i \|^2$$

by Theorem 2.7 and the Parseval identity.]

(b)   Let $C^{(n)}$, $n \geq 1$, be the principal minors of C so that $BC^{(n)} = \Theta^{(n)} - I$. Show that the $C^{(n)}$ converge to C in the Hilbert Schmidt norm. Conclude that $BC^{(n)} = \Theta^{(n)} - I$ converges to $\Theta - I$ in the trace class norm. It follows that the usual, finite dimensional determinant of $\Theta^{(n)}$ converges to the Fredholm determinant of Θ by continuity.

---

[5] Observe that

$$\lim_{s \to 0} \frac{e^s - 1}{\sqrt{|s|}} = 0.$$

6   Spectra

At the beginning of Chapter 3 it was pointed out that the Dirichlet spectrum of a function q in L^2 belongs to the space S of all real, strictly increasing sequences $\sigma = (\sigma_1, \sigma_2, ...)$ of the form

$$\sigma_n = n^2\pi^2 + s + \tilde{\sigma}_n, \qquad n \geq 1,$$

where $s \in \mathbb{R}$ and $\tilde{\sigma} = (\tilde{\sigma}_1, \tilde{\sigma}_2, ...) \in \ell^2$. This observation led us to recast the problem of characterizing spectra as the problem of determining the image of the map μ from L^2 into S.

The principal result of this chapter is that μ maps onto S. In fact, we shall show that μ maps the even subspace E of L^2 onto S.

The strategy here is to construct a special sequence of vectorfields on E. The nth vectorfield is chosen so that the Dirichlet eigenvalues $\mu_j, j \neq n$, are kept fixed by the associated flow while μ_n is moving. In combination, they allow us to shift the spectrum into any desired position, and to reach every point on S.

Such vectorfields are easy to construct. Let $\mathbb{1}_n$, $n \geq 1$, be the constant vectorfield on S given by

$$\mathbb{1}_n = (0, \delta_{mn}, m \geq 1)$$

in the standard coordinate system on S. These vector fields have the desired properties on S. It therefore suffices to pull them back to vectorfields on E

using μ_E. That is, we consider the vectorfields $(d_q\mu_E)^{-1}(\mathbb{1}_n)$. By Corollary 3.1,

$$(d_q\mu_E)^{-1}(\mathbb{1}_n) = -2\frac{d}{dx}(a_n - [a_n]g_n^2)$$

$$= W_n(x, q).$$

More generally, one may consider the vectorfields

$$W_\eta = \eta_0 + \sum_{n \geq 1} \eta_n W_n, \qquad \eta \in \mathbb{R} \times \ell^2$$

on E. In the μ-coordinate system, W_η becomes the constant vectorfield η. Consequently, its solution curves $\phi^t(q, W_\eta)$ become straight lines:

$$\mu(\phi^t(q, W_\eta)) = \mu(q) + t\eta.$$

It follows that $\phi^t(q, W_\eta)$ is a real analytic function of t, η, q and is defined as long as the straight line $\mu(q) + t\eta$ remains inside the open set $\mu(E)$ of S.

So far, the vectorfields W_η and the vectorfields V_ξ discussed in Chapter 4 seem to have similar properties. In contrast to the latter, however, the solution curves of W_η are not defined for all time. For example, the curve $\phi^t(q, W_n)$, $n \geq 1$, is characterized by

$$\mu(\phi^t(q, W_n)) = \mu(q) + t\mathbb{1}_n.$$

Clearly, t is restricted by the condition[1]

$$\mu_{n-1}(q) < \mu_n(q) + t < \mu_{n+1}(q),$$

since $\mu_n + t$ is confined by its neighboring eigenvalues.

Our immediate goal is to show that $\phi^t(q, W_n)$ exists for all t that satisfy this necessary condition. That is, the solution curves exist for the maximal time interval possible. The proof of this important fact is not quite straightforward. One would like to proceed as in Chapter 4 and obtain an a priori bound for $\|\phi^t(q, W_n)\|$ over any closed subset of this maximal interval. This is possible, but requires several additional concepts that have not been developed. Instead, the technique of Chapter 5 is applied, and the solution curves are written down explicitly.

It is convenient to introduce some notation and state a technical lemma. For $n \geq 1$, set

$$w_n(x, \lambda, q) = y_1(x, \lambda) + \frac{y_1(1, \mu_n) - y_1(1, \lambda)}{y_2(1, \lambda)}y_2(x, \lambda).$$

[1] $\mu_0(q) = -\infty$.

This is the unique solution of $-y'' + qy = \lambda y$ satisfying the boundary conditions

$$w_n(0, \lambda) = 1, \qquad w_n(1, \lambda) = y_1(1, \mu_n)$$

for all λ outside the Dirichlet spectrum of q. As a function of λ, w_n has poles at $\lambda = \mu_m$, $m \neq n$, but a removable singularity at $\lambda = \mu_n$, since the limit

$$\lim_{\lambda \to \mu_n} \frac{y_1(1, \mu_n) - y_1(1, \lambda)}{y_2(1, \lambda)} = -\frac{\dot{y}_1(1, \mu_n)}{\dot{y}_2(1, \mu_n)}$$

exists by l'Hospital's rule and Theorem 2.2.

Furthermore, write

$$z_n(x, q) = y_2(x, \mu_n(q), q)$$

for the unnormalized Dirichlet eigenfunctions of q. In this chapter, they are easier to use than the usual normalized eigenfunctions g_n.

Lemma 1. *For each q in L^2, the function*

$$\omega_n(x, \lambda, q) = [w_n, z_n], \qquad n \geq 1$$

is strictly positive on $[0, 1] \times (\mu_{n-1}(q), \mu_{n+1}(q))$.

Proof.[2] The function ω_n is continuous on $[0, 1] \times (\mu_{n-1}, \mu_{n+1})$. Suppose it vanishes, say, on $[0, 1] \times [\mu_n, \mu_{n+1})$. The case of a root in $[0, 1] \times (\mu_{n-1}, \mu_n]$ is handled analogously.

Consider the set

$$\{\lambda \in [\mu_n, \mu_{n+1}): \omega_n > 0 \text{ on } [0, 1] \times [\mu_n, \lambda]\}.$$

This set is not empty, since

$$\omega_n(x, \mu_n) = [y_1, y_2]\Big|_{\lambda = \mu_n} = 1$$

for all $0 \leq x \leq 1$ by the Wronskian identity. Hence it has a least upper bound $\bar{\lambda}$, which is strictly less than μ_{n+1} by assumption.

Now consider the function $\bar{\omega}_n = \omega_n(x, \bar{\lambda})$ on $[0, 1]$.[3] By continuity,

$$\bar{\omega}_n(x) \geq 0, \qquad 0 \leq x \leq 1,$$

---

[2] For helping us with this proof we thank J. Moser and T. Nanda.
[3] Here, $^-$ is a convenient local notation, not to be confused with complex conjugation.

and by construction, $\bar{\omega}_n(\bar{x}) = 0$ for some $\bar{x}$ in $[0, 1]$. However, for all λ, and in particular for $\bar{\lambda}$, we have

$$\omega_n(0, \lambda) = y_1(0, \lambda)z_n'(0) = 1,$$

$$\omega_n(1, \lambda) = y_1(1, \mu_n)z_n'(1) = 1.$$

Hence, $\bar{x}$ must lie in the *interior* of $[0, 1]$. That is, $0 < \bar{x} < 1$.

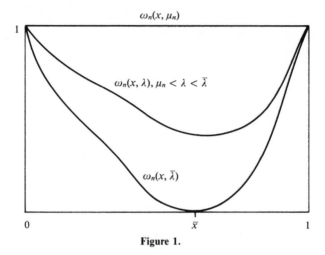

Figure 1.

By the preceding discussion, $\bar{\omega}_n$ has a local minimum at the interior point $\bar{x}$. Hence,

$$0 = \bar{\omega}_n'(\bar{x})$$

$$= (\bar{\lambda} - \mu_n) \cdot \bar{w}_n(\bar{x})z_n(\bar{x}),$$

where $\bar{w}_n = w_n(x, \bar{\lambda})$. Consequently, $\bar{w}_n(\bar{x})$ or $z_n(\bar{x})$ vanishes, since $\bar{\lambda} \neq \mu_n$. But the roots of $\bar{w}_n$ and z_n are all simple, so the equation

$$0 = \bar{\omega}_n(\bar{x})$$

$$= \bar{w}_n z_n' - \bar{w}_n' z_n$$

implies that $\bar{w}_n(\bar{x})$ and $z_n(\bar{x})$ must both vanish. Hence, by Taylor's formula,

$$\bar{w}_n(x) = (x - \bar{x}) \cdot r(x)$$

$$z_n(x) = (x - \bar{x}) \cdot s(x)$$

in a neighborhood of $\bar{x}$ with nonvanishing continuous functions r, s, since $\bar{w}_n$ and z_n are continuously differentiable.[4]

It follows that

$$\bar{\omega}_n'(x) = (\bar{\lambda} - \mu_n) \cdot (x - \bar{x})^2 \cdot r(x)s(x) > 0$$

in a punctured neighborhood of $\bar{x}$. But this implies that $\bar{\omega}_n$ is strictly monotone near $\bar{x}$ and therefore changes sign. This contradicts the fact that $\bar{x}$ is a local minimum of $\bar{\omega}_n$. ■

Theorem 1. *The solution curve $\phi^t(q, W_n)$ of the vectorfield W_n with initial value q in E is given by*

$$\phi^t(q, W_n) = q - 2\frac{d^2}{dx^2} \log \omega_n(x, \mu_n + t, q),$$

and is defined for $\mu_{n-1}(q) < \mu_n(q) + t < \mu_{n+1}(q)$.

The proof of this theorem also provides the unnormalized eigenfunctions along $\phi^t = \phi^t(q, W_n)$. They are

$$z_n(x, \phi^t) = \frac{z_n}{\omega_{n,t}}$$

and

$$z_j(x, \phi^t) = z_j - t\frac{w_{n,t}}{\omega_{n,t}} \int_0^x z_j(s)z_n(s)\, ds, \qquad j \neq n,$$

where $w_{n,t}$, $\omega_{n,t}$ are the functions w_n, ω_n evalutated at $\mu_n + t$, q, and z_j, z_n are evaluated at q.

Proof of Theorem 1. Fix q in E and a positive integer n. Write $w_{n,t}$ and $\omega_{n,t}$ for the functions w_n and ω_n respectively evaluated at $\lambda = \mu_n + t$ and q. The function $w_{n,t}$ is a solution of the equation

(1) $$-y'' + qy = \lambda y$$

for $\lambda = \mu_n + t$, so by Lemma 5.1,

$$h = \frac{1}{z_n}[w_{n,t}, z_n] = \frac{\omega_{n,t}}{z_n}$$

[4] Observe that in general w_n and z_n are not twice continuously differentiable.

is a solution of the equation

(2) $$-y'' + \left(q - 2\frac{d^2}{dx^2}\log z_n\right)y = \lambda y$$

for the same λ. The idea now is to apply Lemma 5.2 to this solution h of equation (2) and the solutions

$$g = z_n, \qquad f = z_j, \qquad j \neq n$$

of equation (1).

By Lemma 1,

$$\omega_{n,t} = gh = z_n h$$

is strictly positive on $[0, 1]$ for $\mu_{n-1} < \mu_n + t < \mu_{n+1}$. Hence,

$$q_t = q - 2\frac{d^2}{dx^2}\log \omega_{n,t}$$

belongs to L^2. So do the functions

$$z_{n,t} = \frac{1}{h} = \frac{z_n}{\omega_{n,t}}$$

and, for $j \neq n$,

$$z_{j,t} = z_j - \frac{1}{\mu_n - \mu_j}\frac{[z_n, z_j]}{z_n}\frac{d}{dx}\log \omega_{n,t}$$

$$= z_j - t\frac{w_{n,t}}{\omega_{n,t}}\frac{[z_j, z_n]}{\mu_j - \mu_n}$$

$$= z_j - t\frac{w_{n,t}}{\omega_{n,t}}\int_0^x z_j z_n \, ds.$$

Here, the identity $[z_j, z_n] = (\mu_j - \mu_n)\int_0^x z_j z_n \, ds$ was used.

Now, by Lemma 5.2, the functions $z_{j,t}, j \geq 1$, are genuine solutions of the equation

$$-y'' + q_t y = \lambda y$$

for $\lambda = \mu_j + t\delta_{jn}$. They vanish at 0 and 1 and have the same number of roots in $[0, 1]$ as $z_{j,0} = z_j$ by Lemma 2.3. In fact, checking the x-derivative at 0 we find that

$$z_{j,t} = z_j(x, q_t)$$

for all $j \geq 1$. Hence,

$$\mu_j(q_t) = \mu_j(q) + t\delta_{jn}, \qquad j \geq 1$$

as required.

It remains to check that the functions κ_j do not change along q_t. Indeed,

$$\kappa_j(q_t) = \log \left| \frac{z'_{j,t}(1)}{z'_{j,t}(0)} \right|$$

$$= \log \left| \frac{z'_j(1)}{z'_j(0)} \right|$$

$$= \kappa_j(q)$$

for all $j \geq 1$, using the fact that $\omega_{n,t} = 1$ at 0 and 1.

It follows that

$$\phi^t(q, W_n) = q_t$$

$$= q - 2 \frac{d^2}{dx^2} \log \omega_{n,t}.$$

In particular, the curve is defined as long as $\mu_{n-1} < \mu_n + t < \mu_{n+1}$. The proof of Theorem 1 is complete. ■

As an illustration some points on the curves $\phi^t(0, W_1)$ and $\phi^t(0, W_2)$ are plotted along with their first normalized eigenfunction.

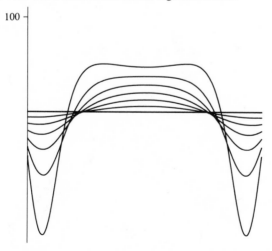

$\phi^t(0, W_1)$ where t increases from 0 so that μ_1 approaches μ_2

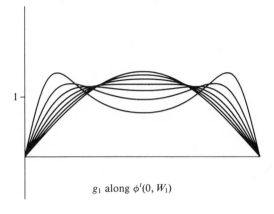

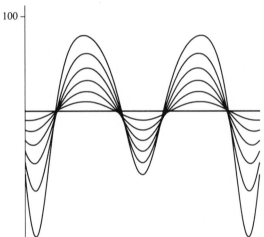

g_1 along $\phi'(0, W_1)$

$\phi'(0, W_2)$ where t increases from 0 so that μ_2 approaches μ_3

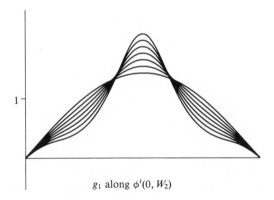

g_1 along $\phi'(0, W_2)$

Problem 1. Show that, with the notation as in the remark following Theorem 1,

$$g_n(x, \phi^t) = \sqrt{\frac{t\dot{y}_2(1, \mu_n)}{y_2(1, \mu_n + t)}} \frac{g_n}{\omega_{n,t}}$$

and

$$g_j(x, \phi^t) = \sqrt{\frac{\mu_n - \mu_j}{\mu_n + t - \mu_j}} \left(g_j + t \frac{w_{n,t}}{\omega_{n,t}} \int_x^1 g_j z_n \, ds \right), \qquad j \neq n.$$

[Hint: Observe that

$$y_2(1, \lambda, \phi^t) = \frac{\mu_n + t - \lambda}{n^2 \pi^2} \prod_{j \neq n} \frac{\mu_j - \lambda}{j^2 \pi^2}$$

and recall Theorem 2.2.]

It is now possible to answer another question raised at the beginning of Chapter 3. Calculating the solution curve of W_1 with initial value $q = 0$ we find that

$$-2 \frac{d^2}{dx^2} \log \left[\frac{\sin \sqrt{\mu_1}(1 - x) - \sin \sqrt{\mu_1} x}{\sin \sqrt{\mu_1}}, \sin \pi x \right]$$

is the unique even function with Dirichlet spectrum

$$\mu_1, 4\pi^2, 9\pi^2, \dots$$

for all μ_1 strictly less than $4\pi^2$. Hence, any such sequence with $\mu_1 < \mu_2(0)$ arises as the Dirichlet spectrum of a function in L^2.

Theorem 2. μ *maps E onto S. Consequently, μ_E is a real analytic isomorphism between E and S.*

Proof. If μ maps E onto S, then μ_E is a real analytic isomorphism between E and S, since μ_E is a local real analytic isomorphism by Theorem 3.2 and is globally one-to-one by Theorem 3.3.

To prove that μ maps onto S, let σ be an arbitrary point in S. In the standard coordinates on S, we have $\sigma = (s, \bar{\sigma})$. It suffices to consider the case $s = 0$. For, if there is a q in E such that

$$\mu(q) = (0, \bar{\sigma}),$$

then

$$\mu(q + s) = (s, \tilde{\sigma}),$$

and $q + s$ is clearly contained in E.

So let $s = 0$. Consider the modified sequences

$$\sigma^N = (\mu_1, ..., \mu_N, \sigma_{N+1}, ...),$$

where $\mu_n = n^2\pi^2$, $1 \le n \le N$. The sequences σ^N converge to $\mu(0)$ as N tends to infinity, so for sufficiently large N, they must be contained in the open image of the map μ. Hence, if N is sufficiently large,

$$(\mu_1, ..., \mu_N, \sigma_{N+1}, ...) = \mu(q)$$

for some q in E. That is, the tale of any point in S with $s = 0$ is attained.

It remains to shift the first N eigenvalues of q to $\mu_1, ..., \mu_N$. This is easily done using the flows of Theorem 1, since only finitely many eigenvalues need to be adjusted.

However, care must be taken to avoid the crossing of eigenvalues. To be on the safe side, first shift $\mu_1, ..., \mu_N$ to the far left, then move them into the desired positions beginning with μ_N.[5]   ∎

We thus have a complete characterization of all possible Dirichlet spectra.

Corollary 1. *The sequence* $\sigma = (\sigma_1, \sigma_2, ...)$ *is the Dirichlet spectrum of some function in* L^2 *if and only if it is real, strictly increasing, and of the form*

$$\sigma_n = n^2\pi^2 + s + \ell^2(n),$$

where s *is a real number.*

Also, Theorem 3.7 can be sharpened:

Corollary 2. $\kappa \times \mu$ *is a real analytic isomorphism between* L^2 *and* $\ell_1^2 \times S$.

There is also a closed formula for

$$\mu_E^{-1}: S \to E.$$

It is the counterpart of the formula for $\exp_q$ which was derived in Chapter 5.

---

[5] More elegantly, one can shift the eigenvalues into position moving each of them only once. See the Proof of Theorem 3.

For the sake of convenience we restrict μ_E^{-1} to the subset

$$S_0 = \{\sigma = (s, \tilde{\sigma}): s = 0\} \subset S.$$

The general case follows from a translation. For σ in S_0 we set

$$\Pi(\sigma) = \prod_{j > i \geq 1} \frac{\sigma_i - j^2\pi^2}{\sigma_i - \sigma_j} \frac{i^2\pi^2 - \sigma_j}{i^2\pi^2 - j^2\pi^2}.$$

The convergence of the infinite product follows from Lemma E.1, since

$$\left| \frac{\sigma_i - j^2\pi^2}{\sigma_i - \sigma_j} - 1 \right| = \left| \frac{\sigma_j - j^2\pi^2}{\sigma_i - \sigma_j} \right| = \frac{l^2(j)}{i^2 - j^2},$$

and similarly for the second factor.

Theorem 3. *For σ in S_0,*

$$\mu_E^{-1}(\sigma) = -2 \frac{d^2}{dx^2} \log(\Pi(\sigma) \det \Omega(x, \sigma)),$$

where Ω is the infinite matrix whose elements are given by

$$\omega_{ij}(x, \sigma) = \frac{\sigma_i - i^2\pi^2}{\sigma_i - j^2\pi^2} \left[\cos\sqrt{\sigma_i}x + \frac{(-1)^i - \cos\sqrt{\sigma_i}}{\sin\sqrt{\sigma_i}} \sin\sqrt{\sigma_i}x, \frac{\sin\pi jx}{\pi j} \right]$$

for $i, j \geq 1$, and the determinant of Ω is defined as in Chapter 5.[6]

Observe that

$$\omega_{ij}(x, \sigma) = \frac{\sigma_i - \mu_i}{\sigma_i - \mu_j} [w_i, z_j] \Big|_{\substack{\lambda = \sigma_i \\ q = 0}},$$

where μ_n, $n \geq 1$, are the Dirichlet eigenvalues of $q = 0$. In particular,

$$\omega_{ii}(x, \sigma) = [w_i, z_i] \Big|_{\substack{\lambda = \sigma_i \\ q = 0}} = \omega_i(x, \sigma, 0)$$

in the notation of Lemma 1.

It is implicit in the statement of the theorem that the product $\Pi(\sigma) \det \Omega(x, \sigma)$ always exists and never vanishes. It is possible that Π vanishes, in which case the determinant of Ω is infinite. This happens

---

[6] This is an unpublished result of J. Ralston and E. Trubowitz.

precisely when $\sigma_i = j^2\pi^2$ for some $i \neq j$. Otherwise, det Ω is finite, does not vanish, and we equivalently have

(3) $$\mu_E^{-1}(\sigma) = -2\frac{d^2}{dx^2}\log \det \Omega(x, \sigma).$$

This formula holds on a relatively open dense subset of S_0. As to the Fredholm determinant of Ω see Problem 3 below.

Proof of Theorem 3. Fix σ in S_0, and let

$$\mu_n = n^2\pi^2, \qquad n \geq 1$$

be the Dirichlet eigenvalues of $q = 0$. The idea is to apply the flows of the vectorfields $W_1, W_2, \ldots$ one after the other to shift $\mu_1, \mu_2, \ldots$ into the positions $\sigma_1, \sigma_2, \ldots$ and to combine the closed expressions provided by Theorem 1. In general, however, we can not move the eigenvalues in their natural succession. For example, if $\sigma_1 > \mu_2$, then we can not begin by moving μ_1 to σ_1, since eigenvalues must not cross. Instead, we have to move them in an order depending on σ as follows.

By the asymptotic behavior of the sequences in S there exists an integer N such that

$$|\mu_n - \sigma_n| < |\mu_m - \sigma_m| \qquad \text{for all } n, m \geq N \text{ and } n \neq m.$$

Hence, if we can move the first N eigenvalues into position, then we can continue to move the remaining eigenvalues in their natural order.

To this end, we show that for any two strings

$$\mu_1 < \mu_2 < \cdots < \mu_n$$

$$\sigma_1 < \sigma_2 < \cdots < \sigma_n$$

of length n there is a way to shift the first sequence into the second preserving order and moving each element once. We show this by induction on n. For strings of length 1 there is nothing to prove. So suppose the statement is true for strings of length $n \geq 1$.

If $\mu_{n+1} < \sigma_{n+1}$, then first shift this element to its new position to get

$$\mu_1 < \mu_2 < \cdots < \mu_n < \sigma_{n+1},$$

then shift the remaining elements using the induction hypotheses. If $\mu_{n+1} \geq \sigma_{n+1}$, interchange these two steps. You first get

$$\sigma_1 < \sigma_2 < \cdots < \sigma_n < \mu_{n+1},$$

then you can move μ_{n+1} to σ_{n+1}. This completes the induction.

In effect there exists a finite permutation $j_1, j_2, \ldots$ of the natural numbers $1, 2, \ldots$ [7] such that the sequences

$$\sigma^n = (\sigma_1^n, \sigma_2^n, \ldots)$$

with

$$\sigma_j^n = \begin{cases} \sigma_j, j \in \{j_1, \ldots, j_n\} \\ \mu_j, j \notin \{j_1, \ldots, j_n\} \end{cases}$$

all belong to S_0.

These sequences σ^n converge to σ in S. Hence the functions

$$p_n = \mu_E^{-1}(\sigma^n), \qquad n \geq 1$$

converge to

$$q = \mu_E^{-1}(\sigma)$$

in L^2 by the continuity of μ_E^{-1}.

On the other hand, we have

$$p_n = \phi^{t_n}(p_{n-1}, W_{j_n}), \qquad t_n = \sigma_{j_n} - \mu_{j_n}$$

with $p_0 = 0$, since p_n and p_{n-1} differ only in their j_nth eigenvalue. Applying Theorem 1 to $p_0, \ldots, p_{n-1}$,

$$p_n = - \sum_{k=1}^{n} 2 \frac{d^2}{dx^2} \log \omega_{j_k}(x, t_k, p_{k-1})$$

$$= -2 \frac{d^2}{dx^2} \log \prod_{k=1}^{n} \omega_{j_k}(x, t_k, p_{k-1}).$$

In Appendix F it is shown that for all sufficiently large n,

(4)
$$\prod_{k=1}^{n} \omega_{j_k}(x, t_k, p_{k-1}) = \Pi^{(n)}(\sigma) \det \Omega^{(n)}(x, \sigma)$$

where

$$\Pi^{(n)}(\sigma) = \prod_{1 \leq i < j \leq n} \frac{\sigma_i - j^2 \pi^2}{\sigma_i - \sigma_j} \frac{i^2 \pi^2 - \sigma_j}{i^2 \pi^2 - j^2 \pi^2},$$

$$\Omega^{(n)}(x, \sigma) = (\omega_{ij})_{1 \leq i, j \leq n},$$

[7] That is, $j_n = n$ for all sufficiently large n.

and the sign is determined by the sign of the permutation $j_1, \ldots, j_n$. Hence,

$$p_n = -2\frac{d^2}{dx^2} \log \Pi^{(n)}(\sigma) \det \Omega^{(n)}(x, \sigma).$$

The left hand side of (4) has value 1 and a vanishing first derivative at $x = 0$. So integrating twice with respect to x and exponentiating we obtain

$$\pm \Pi^{(n)}(\sigma) \det \Omega^{(n)}(x, \sigma) = \exp\left(-\tfrac{1}{2} \int_0^x (x - t)p_n(t)\, dt\right).$$

It follows from the convergence of the p_n that the left hand side converges uniformly on $[0, 1]$ to a positive function. The product $\Pi^{(n)}$ itself converges to Π. So also the determinant of $\Omega^{(n)}$ converges to a finite value, if $\Pi \neq 0$, otherwise it tends to infinity. In any event, going to the limit we obtain

$$q = -2\frac{d^2}{dx^2} \log(\Pi(\sigma) \det \Omega(x, \sigma)). \quad \blacksquare$$

Here are some image points of the map μ_E^{-1}. In the captions, $\mu_n = n^2\pi^2$, $n \geq 1$, are the eigenvalues of $q = 0$.

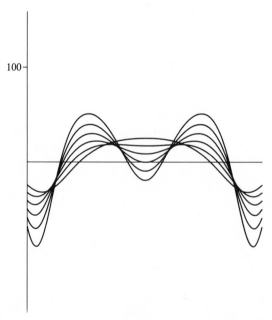

σ_1 is fixed at a value between μ_1 and μ_2, while σ_2 tends towards μ_3; otherwise, $\sigma_n = \mu_n$

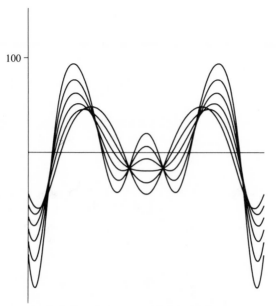

As before, but now σ_3 tends towards μ_4

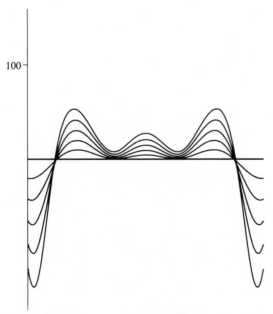

$\sigma_1, \sigma_2, \sigma_3$ tend towards μ_2, μ_3, μ_4 respectively at different rates; otherwise, $\sigma_n = \mu_n$

We mention a natural generalization of the preceding results. Consider the sets

$$K(p) = \kappa^{-1}(\kappa(p))$$

$$= \{q \in L^2 : \kappa(q) = \kappa(p)\}$$

of all functions in L^2 with the same κ-values as a given function p. As a special case we have

$$K(0) = E$$

by Lemma 3.4. These sets are counterparts of the isospectral sets $M(p)$.

$K(p)$ may be viewed as a vertical slice of the real analytic coordinate system $\kappa \times \mu$ on L^2. So each such set is a real analytic submanifold of L^2, and μ is a global coordinate system on it mapping into $\mathbb{R} \times \ell^2$.

Problem 2. (a) Show that with q also the straight line $q + c$, $-\infty < c < \infty$, is contained in $K(p)$.
[Hint: Use that $y_2(x, \lambda, q + c) = y_2(x, \lambda - c, q)$.]
(b) Show that

$$T_q K(p) = \left\{ W_\eta = \eta_0 + \sum_{n \geq 1} \eta_n W_n : \eta \in \mathbb{R} \times \ell^2 \right\}.$$

(c) Show that the derivative of μ is a linear isomorphism between $T_q K(p)$ and $T_{\mu(q)} S \simeq \mathbb{R} \times \ell^2$ given by

$$d_q \mu(W_\eta) = \eta.$$

Theorem 2 generalizes to

Theorem 2*. *For all p in L^2 the restriction $\mu_{K(p)}$ of μ to $K(p)$ is a real analytic isomorphism between $K(p)$ and S.*

Making use of the last Problem the proof is essentially the same as for Theorem 2. Observe that the vectorfields W_n are tangent to $K(p)$, so any solution curve with initial value in $K(p)$ stays in that set.

There is also a closed formula for $\mu_{K(p)}^{-1}$. Restricting this map to the subset

$$S_p = \{\sigma = (s, \tilde{\sigma}) : s = [p]\} \subset S$$

it is given by

$$\mu_{K(p)}^{-1} = p - 2\frac{d^2}{dx^2}\log(\Pi(\sigma, p) \det \Omega(x, \sigma, p)),$$

where

$$\Pi(\sigma, p) = \prod_{j > i \geq 1} \frac{\sigma_i - \mu_j}{\sigma_i - \sigma_j} \frac{\mu_i - \sigma_j}{\mu_i - \mu_j},$$

Ω is the infinite matrix with elements

$$\omega_{ij}(x, \sigma, p) = \frac{\sigma_i - \mu_i}{\sigma_i - \mu_j} [w_i, z_j] \Bigg|_{\substack{\lambda = \sigma_i \\ q = p}},$$

and μ_n, $n \geq 1$, are the Dirichlet eigenvalues of p. The proof is the same as that of Theorem 3.

Problem 3 (The Fredholm determinant of Ω). Fix p in L^2, let μ_n, $n \geq 1$, be its Dirichlet eigenvalues, and let σ be in S_p. For the notion of the Fredholm determinant and some related facts, see Problem 5.4.

(a) Verify the identity

$$\omega_{ij}(x, \sigma, p) = \frac{\sigma_i - \mu_i}{\sigma_i - \mu_j} + (\sigma_i - \mu_i) \int_0^x w_i(s, \sigma_i) z_j(s) \, ds.$$

Show that the three matrices

$$((\sigma_i - \mu_i)\,\delta_{ij}), \quad \left(\frac{1 - \delta_{ij}}{\sigma_i - \mu_j}\right), \quad \left(\int_0^x w_i(s, \sigma_i) z_j(s) \, ds\right)$$

define Hilbert-Schmidt operators on ℓ^2, provided that $\sigma_i \neq \mu_j$ for $i \neq j$. Conclude that in this case, $\Omega - I$ is trace class on ℓ^2, hence the Fredholm determinant of Ω is well defined.

(b) Imitate the argument of Problem 5.3 to show that the finite dimensional determinants of $\Omega^{(n)}$ converge to the Fredholm determinant of Ω as $n \to \infty$.

A Continuity, Differentiability and Analyticity

We discuss the notions of continuity, differentiability and especially analyticity for maps between Banach spaces.

Let U be an arbitrary subset of a Banach space E, and let F be another Banach space. A map

$$f: U \to F$$

is *continuous* on U, if it maps strongly convergent sequences in U into strongly convergent sequences in F. That is, if $x_n \to x$ strongly in U, then $f(x_n) \to f(x)$ strongly in F.

This is the familiar notion of a continuous map. However, in an infinite dimensional setting, continuous maps lack many of the useful properties one would like to have. For instance, they need not be bounded on closed bounded subsets of E. This is due to the fact that the closed unit ball is no longer compact in the strong topology.

Problem 1. Construct a continuous, unbounded function with bounded support.

[Hint: Choose a sequence of unit vectors x_n such that $\|x_n - x_m\| \geq \delta > 0$ for $m \neq n$. For each n, construct a function f_n using Urysohn's lemma which is equal to n at x_n and vanishes outside $\|x - x_n\| \geq \delta/3$. Add these functions together.]

A stronger notion of continuity, which rules out such behavior, is obtained by generalizing the notion of a compact operator familiar from linear functional analysis. A map $f: U \to F$ is *compact* on U, if it maps weakly convergent sequences in U into strongly convergent sequences in F. That is, if $x_n \to x$ weakly in U, then $f(x_n) \to f(x)$ strongly in F.

A strongly convergent sequence is also weakly convergent, so every compact map is continuous. But not every continuous map is compact.

Problem 2. Show that the function

$$u \to \tfrac{1}{2} \int_0^1 u^2(t) \, dt$$

is continuous on $L^2_\mathbb{C}[0, 1]$, but not compact.

The closed unit ball, and more generally every closed, bounded and convex subset of a reflexive Banach space is weakly compact regardless of the dimension. For this reason, compact maps are much nicer than merely continuous maps. For instance, the norm of a compact function on a closed, bounded, convex subset is bounded and attains its maximum. Moreover, the function itself is uniformly continuous.

We now turn to differentiable maps. Let $U \subset E$ be open. As usual, the map

$$f: U \to F$$

is *differentiable at* $x \in U$, if there exists a bounded linear map

$$d_x f: E \to F$$

such that

$$\|f(x + h) - f(x) - d_x f(h)\| = o(\|h\|).$$

That is, for every $\varepsilon > 0$ there is a $\delta > 0$ such that $\|f(x + h) - f(x) - d_x f(h)\| \leq \varepsilon \|h\|$ for all h with $\|h\| < \delta$. The linear map $d_x f$ is uniquely determined, and is called the *derivative of f at x.*

The map f is differentiable on U, if it is differentiable at each point in U. In this case, the derivative is a map from U into the Banach space $L(E, F)$ of all bounded linear maps from E into F, denoted by df. If this map is continuous, then f is *continuously differentiable,* or of class C^1, on U.

We frequently encounter complex valued differentiable functions f, which are defined on open subsets of $L_{\mathbb{C}}^2[0, 1]$. In this case, $d_q f$ is a bounded linear functional on $L_{\mathbb{C}}^2[0, 1]$. It follows from the Riesz representation theorem that there is a unique element $\partial f/\partial q$ in $L_{\mathbb{C}}^2[0, 1]$ such that

$$d_q f(v) = \left\langle v, \frac{\overline{\partial f}}{\partial q} \right\rangle = \int_0^1 v(t) \frac{\partial f}{\partial q(t)} \, dt$$

for all $v \in L_{\mathbb{C}}^2[0, 1]$. The function $\partial f/\partial q$ is the *gradient* of f at q. Our notation imitates the usual finite dimensional notation.[1]

Problem 3. Suppose f is differentiable on $U \subset E$. Show that

$$d_x f(h) = \frac{d}{d\varepsilon} f(x + \varepsilon h)\Big|_0$$

$$= \lim_{\varepsilon \to 0} \frac{1}{\varepsilon} (f(x + \varepsilon h) - f(x))$$

for all $x \in U$ and $h \in E$. The right hand side is called the *directional derivative of f at x in the direction h.*

Conversely, suppose the limit

$$\delta_x(h) = \lim_{\varepsilon \to 0} \frac{1}{\varepsilon} (f(x + \varepsilon h) - f(x))$$

exists locally uniformly in x and h. That is, suppose for each x in U and $\delta > 0$ there exists a neighborhood V of x and an $\varepsilon_0 > 0$ such that

$$\left\| \frac{1}{\varepsilon} (f(\xi + \varepsilon h) - f(\xi)) - \delta_\xi(h) \right\| < \delta$$

for all ξ in V, $\|h\| < 1$ and $|\varepsilon| < \varepsilon_0$. Show that f is continuously differentiable on U, and that $d_x f = \delta_x$.

[Hint: To prove additivity of δ_x in h write

$$\delta_x(h + k) - \delta_x(h) - \delta_x(k)$$

$$= \lim_{\varepsilon \to 0} \frac{1}{\varepsilon} (f(x + \varepsilon h + \varepsilon k) + f(x) - f(x + \varepsilon h) - f(x + \varepsilon k)).$$

Add and subtract terms to rewrite the last expression as a combination of four different difference quotients at the point $\xi = x + \varepsilon(h + k)/2$.]

---

[1] The gradient is defined in the same way for the complexification of any real Hilbert space.

Here are several examples of differentiable maps.

Example 1.   A bounded linear map $L: E \to F$ is continuously differentiable on E, and $d_x L = L$ for all x.

Example 2.   The function $f: q \to \int_0^1 q(t)\, dt$ is continuously differentiable on $L_{\mathbb{C}}^2[0, 1]$. Its derivative is given by

$$d_q f(v) = \int_0^1 v(t)\, dt = \langle v, 1 \rangle,$$

hence $\partial f / \partial q = 1$.

Example 3.   The function $g: q \to \frac{1}{2} \int_0^1 q^2(t)\, dt$ is also C^1 on $L_{\mathbb{C}}^2[0, 1]$, its derivative is

$$d_q g(v) = \int_0^1 v(t) q(t)\, dt = \langle v, \bar{q} \rangle.$$

Consequently, $\partial g / \partial q = q$.

Example 4.   Let $M(n, \mathbb{C})$ be the Hilbert space of all complex $n \times n$-matrices with inner product $\langle A, B \rangle = \sum_{i,j} a_{ij} \bar{b}_{ij}$, where a_{ij}, b_{ij} are the entries of A and B respectively. The determinant function

$$\det: A \to \det(A)$$

is C^1 on $M(n, \mathbb{C})$, since it is a polynomial in the entries of A. The gradient of det at A is the adjoint of A,

$$\frac{\partial \det}{\partial A} = \mathrm{adj}(A),$$

which is the matrix of cofactors $a^{ij} = (-1)^{i+j} \det(a_{kl})_{k \neq j, l \neq i}$ of A. This may also be written as

$$\frac{\partial \det}{\partial A} = \det(A) A^{-1}.$$

in case A is invertible.

For the proof, let $1 \leq j \leq n$, and apply $d_A \det$ to a matrix U, which is nonzero only in the jth column. Expanding the determinant with respect to this column you get

$$d_A \det(U) = \frac{d}{d\varepsilon} \det(A + \varepsilon U) \bigg|_{\varepsilon = 0}$$

$$= \frac{d}{d\varepsilon} \left(\sum_{i=1}^{n} (a_{ij} + \varepsilon u_{ij}) a^{ij} \right) \bigg|_{\varepsilon = 0}$$

$$= \sum_{i=1}^{n} u_{ij} a^{ij}$$

$$= \langle U, \overline{\mathrm{adj}(A)} \rangle.$$

By linearity, this holds also for general U. Hence, $\partial \det / \partial A = \mathrm{adj}(A)$.

Problem 4. Let A_j, $1 \leq j \leq n$, denote the jth column of a matrix A. Show that the directional derivative of det in the direction U is given by

$$d_A \det(U) = \sum_{j=1}^{n} \det(A_1, \ldots, U_j, \ldots, A_n)$$

$$= \det(A) \, \mathrm{tr}(A^{-1} U),$$

where the last term applies only to invertible matrices A.

Example 5. Let α be a 1-form on $\mathbb{R}^n$ with smooth coefficients, and let $C^1(T^1, \mathbb{R}^n)$ be the space of all continuously differentiable loops in $\mathbb{R}^n$, that is, all periodic continuously differentiable maps from the real line into $\mathbb{R}^n$ with period 1. Then the map

$$\mathrm{loop}_\alpha : \gamma \to \int_\gamma \alpha = \int_0^1 \gamma^* \alpha$$

is continuously differentiable on $C^1(T^1, \mathbb{R}^n)$, and its derivative is the linear map

$$d_\gamma \, \mathrm{loop}_\alpha : \psi \to \int_\gamma i_\psi \, d\alpha,$$

where i_ψ denotes interior multiplication by $\psi : i_\psi d\alpha = d\alpha(\psi, \cdot)$.

This can be seen as follows. By the calculus of differential forms, or by a direct computation using coordinates,

$$(\gamma + \psi)^*\alpha - \gamma^*\alpha = \int_0^1 \frac{d}{ds}(\gamma + s\psi)^*\alpha \, ds$$

$$= \int_0^1 (\gamma + s\psi)^*(i_\psi d\alpha + d i_\psi \alpha) \, ds.$$

The integral of $(\gamma + s\psi)^* d i_\psi \alpha$ over $[0, 1]$ vanishes by periodicity. It follows that

$$\text{loop}_\alpha(\gamma + \psi) - \text{loop}_\alpha(\gamma) - \int_\gamma i_\psi d\alpha = \int_0^1 [(\gamma + \psi)^*\alpha - \gamma^*\alpha - \gamma^* i_\psi \, d\alpha]$$

$$= \int_0^1 \int_0^1 [(\gamma + s\psi)^* i_\psi \, d\alpha - \gamma^* i_\psi \, d\alpha] \, ds.$$

The last term is $O(\|\psi\|_{C^1}^2)$.

Higher derivatives are defined inductively. If $f: U \to F$ is differentiable on U, and if its derivative $df: U \to L(E, F)$ is also differentiable on U, then f is twice differentiable. Its second derivative is a map

$$d^2 f: U \to L^2(E, F)$$

from U into the Banach space $L^2(E, F)$ of all bounded, bilinear maps from $E \times E$ into F. The map is of class C^2, if $d^2 f$ is continuous.

In general, f is p times differentiable on U, $p \geq 1$, if

$$d^p f = d(d^{p-1}f): U \to L^p(E, F)$$

exists as a map from U into the Banach space $L^p(E, F)$ of all bounded p-linear maps from $E \times \cdots \times E$ (p times) into F. It is of class C^p, if $d^p f$ is continuous.

Finally, f is *smooth*, or of class C^∞, if it is of class C^p for all $p \geq 1$.

If f is C^p, then actually $d_x^p f$ is a *symmetric* bounded p-linear map from $E \times \cdots \times E$ into F. This is a restatement of the classical theorem about interchanging the order of partial differentiation. For a proof, see for instance Dieudonné [Di] or Lang [La1].

We are not going to list and prove all the elementary properties of differentiable maps and their derivatives. They can be found in the standard references. However, because it is used so often, we mention the chain rule.

If $f: U \to V$ and $g: V \to W$ are two C^1-maps, then the composite map $g \circ f: U \to W$ is also C^1, and

$$d_x(g \circ f) = d_{f(x)}g \cdot d_x f$$

is its derivative at x.

Let us also recall Taylor's formula with integral remainder. If f is p times continuously differentiable, $p \geq 1$, and if the segment $x + th$, $0 \leq t \leq 1$, is contained in U, then

$$f(x + h) = f(x) + d_x f(h) + \cdots + \frac{1}{(p-1)!} d_x^{p-1} f(k, \ldots, h)$$

$$+ \int_0^1 \frac{(1-t)^{p-1}}{(p-1)!} d_{x+th}^p f(h, \ldots, h) \, dt.$$

The integral remainder is bounded by $(\|h\|^p / p!) \sup_{0 \leq t \leq 1} \|d_{x+th}^p f\|.$[2] A special case is the "mean value theorem", where $p = 1$:

$$\|f(x) - f(y)\| \leq \|x - y\| \sup_{0 \leq t \leq 1} \|d_{tx+(1-t)y} f\|.$$

If, in addition, f is smooth, and the integral remainder converges to 0 as p tends to infinity uniformly in some ball $\|h\| < r$, then f admits a *Taylor series expansion*

$$f(x + h) = \sum_{k \geq 0} \frac{1}{k!} d_x^k f(h, \ldots, h)$$

at x in this ball.

The integral in Taylor's formula is the familiar Riemann integral of a continuous, Banach space valued function on $[0, 1]$. To recall its definition, let G be a Banach space. For a step map $s: [0, 1] \to G$, one sets

$$\int_0^1 s(t) \, dt = \sum_{i=1}^n (t_i - t_{i-1}) \cdot s_i$$

where $0 = t_0 < t_1 < \cdots < t_n = 1$ is a partition of $[0, 1]$ such that $s(t) = s_i$ for $t_{i-1} < t < t_i$. For two step maps s and s',

$$\left\| \int_0^1 s(t) \, dt - \int_0^1 s'(t) \, dt \right\| \leq \sup_{0 \leq t \leq 1} \|s(t) - s'(t)\|.$$

---

[2] Here and in similar cases, $\|d^p f\|$ is understood to denote the operator norm of $d^p f$. That is, for L in $L^p(E, F)$,

$$\|L\| = \sup_{\|h_i\| \leq 1, \, 1 \leq i \leq n} \|L(h_1, \ldots, h_n)\|.$$

It follows that

$$\int_0^1 g(t)\, dt = \lim_{m \to \infty} \int_0^1 s_m(t)\, dt$$

is well defined for any map $g: [0, 1] \to G$, which is the uniform limit of step maps s_m. In particular, the integral is defined for continuous maps g.

All the usual properties of an integral hold. For example,

$$\left\| \int_0^1 g(t)\, dt \right\| \le \int_0^1 \|g(t)\|\, dt,$$

and

$$L \int_0^1 g(t)\, dt = \int_0^1 Lg(t)\, dt$$

for every $L \in G^*$, the dual space of G.

The multiple Riemann integral of continuous Banach space valued maps on $[0, 1]^n$ is constructed in the same way. The order of integration may be interchanged.

We finally turn to analytic maps. Suppose

$$f: U \to F$$

is a map from an open subset U of a *complex* Banach space E into a complex Banach space F. By definition, f is *analytic* on U, if it is continuously differentiable on U. This is the straightforward generalization of Riemann's notion of an analytic function of one complex variable.

Example 6. Let $n \ge 1$, and $A \in L_s^n(E, F)$, the Banach space of all bounded, n-linear symmetric maps from $E \times \cdots \times E$ into F. Then

$$P(x) = A(x, \ldots, x),$$

the evaluation of A on the diagonal, is analytic on E. The function P is called a homogeneous, F-valued polynomial of degree n on E. We also use the notation $P = \hat{A}$ to indicate the linear map A from which P is obtained.

Example 7. Let P_n, $n \ge 0$, be homogeneous polynomials of degree n on E, where P_0 is simply a constant. If the "power series"

$$\sum_{n \ge 0} P_n(x)$$

converges absolutely and uniformly on an open subset $U \subset E$, then it defines an analytic function on U.

Problem 5. (a) Let $A \in L_s^n(E, F)$, and let $P = \hat{A}$ be the corresponding homogeneous polynomial. Verify the polarization identity

$$A(x_1, \ldots, x_n) = \frac{1}{2^n n!} \sum_{\substack{\varepsilon_i = \pm 1 \\ 1 \leq i \leq n}} \varepsilon_1 \cdots \cdot \varepsilon_n P(\varepsilon_1 x_1 + \cdots + \varepsilon_n x_n).$$

(b) Show that there is a one-to-one correspondence between bounded symmetric n-linear maps on E and homogeneous polynomials of degree n on E.

It is convenient to introduce another notion of analyticity. The map $f: U \to F$ is *weakly analytic on U*, if for each $x \in U$, $h \in E$ and $L \in F^*$, the function

$$z \to Lf(x + zh)$$

is analytic in some neighborhood of the origin in $\mathbb{C}$ in the usual sense of one complex variable. The *radius of weak analyticity of f at x* is the supremum of all $r \geq 0$ such that the above function is defined and analytic in the disc $|z| < 1$ for all $L \in F^*$ and $h \in E$ with $\|h\| < r$.

It is easy to see that the radius r of weak analyticity at x is equal to the distance ρ of x to the boundary of U. For, $r \leq \rho$ by definition. On the other hand, if L and h are given with $\|h\| < \rho$, then the function $z \to Lf(x + zh)$ is well defined on the disc $|z| < 1$ and analytic in some neighborhood of each point in it, since f is weakly analytic on all of U. Consequently, this function is analytic on $|z| < 1$, and so also $r \geq \rho$. It follows that $r = \rho$.

The notion of a weakly analytic map is weaker than that of an analytic map. For instance, every globally defined, but unbounded linear operator is weakly analytic, but not analytic. Remarkably, a weakly analytic map is analytic, if, in addition, it is locally bounded, that is, bounded in some neighborhood of each point of its domain of definition.

Theorem 1. *Let $f: U \to F$ be a map from an open subset U of a complex Banach space E into a complex Banach space F. Then the following three statements are equivalent.*

(1) *f is analytic on U.*
(2) *f is locally bounded and weakly analytic on U.*
(3) *f is infinitely often differentiable on U, and is represented by its Taylor series in a neighborhood of each point in U.*

A prerequisite for the proof is a version of

Cauchy's Formula. *Suppose f is weakly analytic and continuous on U. Then, for every $x \in U$ and $h \in E$,*

$$f(x + zh) = \frac{1}{2\pi i} \int_{|\zeta| = \rho} \frac{f(x + \zeta h)}{\zeta - z} d\zeta, \qquad |z| < \rho < r/\|h\|,$$

where r is the radius of weak analyticity of f at x.

Proof. Fix x in U, and let $r > 0$ be the radius of weak analyticity of f at x. Then the open ball of radius r around x is contained in U. For every $h \in E$, the integral

$$\frac{1}{2\pi i} \int_{|\zeta| = \rho} \frac{f(x + \zeta h)}{\zeta - z} d\zeta, \qquad |z| < \rho < r/\|h\|$$

is well defined, since f is continuous and $\|\zeta h\| < r$, $|\zeta - z| > 0$ for $|\zeta| = \rho$. Then, for every $L \in F^*$,

$$\frac{1}{2\pi i} L \int_{|\zeta| = \rho} \frac{f(x + \zeta h)}{\zeta - z} d\zeta = \frac{1}{2\pi i} \int_{|\zeta| = \rho} \frac{Lf(x + \zeta h)}{\zeta - z} d\zeta$$

$$= Lf(x + zh)$$

by the usual Cauchy formula. Since this holds for all L, the statement follows. ■

Proof of Theorem 1. (1) $\Rightarrow$ (2) Suppose f is analytic. A differentiable map is continuous, and a continuous map is locally bounded, so f is locally bounded. Furthermore, for every $L \in F^*$, the composite map $Lf(x + zh)$ is continuously differentiable in z wherever it is defined. Thus, f is weakly analytic.

(2) $\Rightarrow$ (1) Suppose f is locally bounded and weakly analytic. We first show that f is continuous.

Fix $x \in U$ and choose $r > 0$ so small that

$$\sup_{\|h\| \le r} \|f(x + h)\| = M < \infty.$$

By the usual Cauchy formula,

$$Lf(x + zh) - Lf(x) = \frac{1}{2\pi i} \int_{|\zeta| = 1} z \frac{Lf(x + \zeta h)}{(\zeta - z)\zeta} d\zeta$$

for $|z| < 1$, $\|h\| < r$. Hence, for $|z| < \frac{1}{2}$,

$$\left| \frac{Lf(x + zh) - Lf(x)}{z} \right| \leq 2M\|L\|,$$

where $\|L\|$ denotes the operator norm of L. This estimate holds for all L in F^* uniformly for $|z| < \frac{1}{2}$ and $\|h\| < r$. Consequently,

$$\left\| \frac{f(x + zh) - f(x)}{z} \right\| \leq 2M$$

for $|z| < \frac{1}{2}$ and $\|h\| < r$. From this, the continuity of f follows.

Now, f being weakly analytic and continuous, Cauchy's formula applies, and

$$f(x + zh) = \frac{1}{2\pi i} \int_{|\zeta| = 1} \frac{f(x + \zeta h)}{\zeta - z} d\zeta$$

for $|z| < 1$ and $\|h\| < r$. It follows that f has directional derivatives in every direction h, namely

$$\delta_x(h) = \lim_{z \to 0} \frac{1}{z} (f(x + zh) - f(x))$$

$$= \frac{1}{2\pi i} \int_{|\zeta| = 1} \frac{f(x + \zeta h)}{\zeta^2} d\zeta.$$

In fact, this limit is uniform in $\|\xi - x\| < r/2$ and $\|h\| < r/2$, since

$$\left\| \frac{1}{z} (f(\xi + zh) - f(\xi)) - \delta_\xi(h) \right\| = \left\| \frac{1}{2\pi i} \int_{|\zeta| = 1} z \frac{f(x + \zeta h)}{\zeta^2(\zeta - z)} d\zeta \right\|$$

$$\leq 2M|z|$$

for $|z| < \frac{1}{2}$. It follows from Problem 3, that f is continuously differentiable, hence analytic on U.

(1) $\Rightarrow$ (3) Suppose f is analytic on U. Fix x in U and $r > 0$ such that

$$\sup_{\|h\| \leq r} \|f(x + h)\| = M < \infty.$$

For h in E and $n \geq 0$, define

$$P_n(h) = \frac{n!}{2\pi i} \int_{|\zeta| = \rho} \frac{f(x + \zeta h)}{\zeta^{n+1}} d\zeta,$$

where $\rho > 0$ is chosen sufficiently small. The integral is independent of ρ as long as $0 < \rho \leq r/\|h\|$, since f is analytic.

For instance, $P_0(h) = f(x)$ and $P_1(h) = d_x f(h)$. We show that $P_n(h)$ is the nth directional derivative of f in the direction h.

First of all, Cauchy's formula and the expansion

$$\frac{1}{\zeta - 1} = \sum_{n=0}^{p} \frac{1}{\zeta^{n+1}} + \frac{1}{\zeta^{p+1}(\zeta - 1)}$$

yield

$$f(x + h) - \sum_{n=0}^{p} \frac{1}{n!} P_n(h) = \frac{1}{2\pi i} \int_{|\zeta| = \rho} \frac{f(x + \zeta h)}{\zeta^{p+1}(\zeta - 1)} d\zeta$$

for $\|h\| < r$. Choosing $\rho = r/\|h\|$ for $h \neq 0$, the norm of the right hand side is bounded by

$$M\left(\frac{\|h\|}{r}\right)^p \frac{\|h\|}{r - \|h\|} = \frac{M}{r^p(r - \|h\|)} \|h\|^{p+1}.$$

Consequently,

$$f(x + h) = \sum_{n=0}^{\infty} \frac{1}{n!} P_n(h)$$

for $\|h\| < r$. Moreover, the sum converges uniformly in every ball $\|h\| < \rho < r$.

We now show that each P_n is a homogeneous polynomial of degree n. That is, there exists a bounded symmetric n-linear map A_n such that $P_n = \hat{A}_n$, the polynomial associated to A_n (see Example 6).

Consider the map A_n defined by

$$A_n(h_1, \ldots, h_n)$$

$$= \left(\frac{1}{2\pi i}\right)^n \int_{|\zeta_1| = \varepsilon} \cdots \int_{|\zeta_n| = \varepsilon} \frac{f(x + \zeta_1 h_1 + \cdots + \zeta_n h_n)}{\zeta_1^2 \cdots \zeta_n^2} d\zeta_1 \cdots d\zeta_n,$$

where $\varepsilon > 0$ is sufficiently small, say $\varepsilon < \min r/\|h_i\|$. For every $L \in F^*$, the map

$$(z_1, \ldots, z_n) \to Lf(x + z_1 h_1 + \cdots + z_n h_n)$$

is analytic in a neighborhood of the origin in $\mathbb{C}^n$. Hence, by the usual Cauchy formula for n complex variables,

$$LA_n(h_1, \ldots, h_n) = \frac{1}{dz_1} \cdots \frac{1}{dz_n} Lf(x + z_1 h_1 + \cdots + z_n h_n)\bigg|_0.$$

It follows that A_n is linear and symmetric in all arguments. A_n is also bounded by a straightforward estimate. Finally, using Cauchy's formula again,

$$LA_n(h, \ldots, h) = \left(\frac{d}{dz}\right)^n Lf(x + zh)\bigg|_0$$

$$= LP_n(h)$$

for all L, and therefore

$$A_n(h, \ldots, h) = \hat{A}_n(h) = P_n(h),$$

as we wanted to show.

Thus, on the ball of radius r around x, the map f is represented by a power series, which converges uniformly on every smaller ball around x. It is a basic fact that such a map is infinitely often differentiable. In particular, we have

$$d_x^n f = A_n$$

for all $n \geq 0$.

(3) $\Rightarrow$ (1) is trivial. ∎

Problem 6.   Let c_0 be the Banach space of all sequences $z = (z_1, z_2, \ldots)$ of complex numbers tending to zero, normed by $|z| = \sup_{n \geq 1} |z_n|$. For $n \geq 1$, define P_n on c_0 by

$$P_n(z) = z_1 \cdots z_n.$$

(a)   Show that each P_n is a homogeneous polynomial of degree n on c_0.

(b)   Show that

$$\sum_{n \geq 1} P_n(z)$$

converges uniformly on every ball with a radius smaller than 1 to a function F which is analytic on c_0.

(c)   Show that F is unbounded on every closed ball of radius 1.

Theorem 2.   *Let f_n be a sequence of analytic maps on $U \subset E$, which converges uniformly to a map f. Then f is also analytic.*

Proof.   Given x in U, choose $r > 0$ so that the sequence f_n converges uniformly on the ball of radius r around x. Then, for every $L \in F^*$ and every $h \in E$, $\|h\| < r$, the function

$$z \rightarrow Lf(x + zh)$$

is, on $|z| < 1$, the uniform limit of the analytic functions $z \to Lf_n(x + zh)$, hence is itself analytic. It follows that f is weakly analytic on U. The limit is locally bounded, since it is continuous. By Theorem 1, f is analytic on U.   ∎

It is clear that a map from an open subset of $\mathbb{C}^n$ into $\mathbb{C}^m$ is analytic if and only if all its m coordinate functions are analytic functions of their n arguments. This is not necessarily so in infinite dimensions, because a map may be discontinuous although every coordinate function is analytic. An additional assumption is required.

Theorem 3. *Let $f: U \to H$ be a map from an open subset U of a complex Banach space into a Hilbert space with orthonormal basis e_n, $n \geq 1$. Then f is analytic on U if and only if f is locally bounded, and each "coordinate function"*

$$f_n = \langle f, e_n \rangle: U \to \mathbb{C}$$

is analytic on U. Moreover, the derivative of f is given by the derivatives of its "coordinate functions":

$$df(h) = \sum_{n \geq 1} df_n(h)e_n.$$

Proof. Let $L \in H^*$. By the Riesz representation theorem, there is a unique element ℓ in H such that $L\phi = \langle \phi, \ell \rangle$ for all ϕ in H. Write

$$\ell = \sum_{n \geq 1} \lambda_n e_n,$$

and set

$$\ell_m = \sum_{n=1}^{m} \lambda_n e_n, \qquad m \geq 1.$$

Then L is the operator norm limit of the functionals L_m defined by - $L_m \phi = \langle \phi, \ell_m \rangle$. That is,

$$\sup_{\|\phi\| \geq 1} \|(L - L_m)(\phi)\| \to 0$$

as $m \to \infty$.

Now, given x in U, choose $r > 0$ so that f is bounded on the ball of radius r around x. Fix h in the complex Banach space containing U with $\|h\| < r$.

On $|z| < 1$, the functions

$$z \to L_m f(x + zh) = \sum_{n=1}^{m} \lambda_n f_n(x + zh), \qquad m \geq 1$$

are analytic by hypotheses and tend uniformly to the function

$$z \to Lf(x + zh),$$

since f is bounded. Hence that function is also analytic on $|z| < 1$. This shows that f is weakly analytic and locally bounded. By Theorem 1, the function f is analytic.

Conversely, if f is analytic, then of course, f is locally bounded, and each coordinate function is analytic.

Finally, if f is analytic, then $d_x f(h)$ exists and is an element of H, hence can be expanded with respect to the orthonormal basis e_n, $n \geq 1$. Its nth coefficient is

$$\langle d_x f(h), e_n \rangle = d_x \langle f, e_n \rangle(h) = d_x f_n(h)$$

by the chain rule, since $\langle \cdot, e_n \rangle$ is a linear function. Thus

$$d_x f(h) = \sum_{n \geq 1} d_x f_n(h) e_n$$

as was to be proven. ∎

Finally, we introduce the notion of a real analytic map. Let E, F be real Banach spaces, let $\mathbb{C}E$, $\mathbb{C}F$ be their complexifications, and let $U \subset E$ be open. A map

$$f \colon U \to F$$

is *real analytic* on U, if for each point in U there is a neighborhood $V \subset \mathbb{C}E$ and an analytic map

$$g \colon V \to \mathbb{C}F,$$

such that

$$f = g \qquad \text{on} \qquad U \cap V.$$

It follows that a real analytic map can be expanded into a Taylor series with real coefficients in a ball at each point. The converse is also true.

For more information about analytic maps, we refer the reader for example to [Na] and [Din].

B   Some Calculus

In this appendix the inverse and implicit function theorems are derived in the framework of Banach spaces of arbitrary dimension. Of course, these theorems are well known and can be found in every standard text book on analysis. They are included here because we use them so often. Also, the basic existence and uniqueness theorem for solution curves of vectorfields is reviewed.

We will restrict ourselves to the analytic case, since this suffices for our applications. Also, the proofs are slightly simpler, since we can use the fact that the locally uniform limit of analytic maps is analytic. See Theorem A.2.

First some terminology. Let E, F be Banach spaces and U open in E. The map

$$f: U \to F$$

is a (*real*) *analytic isomorphism on* U, if it is a homeomorphism between U and its image V in F, and f, f^{-1} are (real) analytic on U, V respectively. The map f is a *local* (*real*) *analytic isomorphism at* a in U, if its restriction to some neighborhood A of a is a [real] analytic isomorphism on A.

Inverse Function Theorem.   *Let $f: U \to F$ be an analytic map from an open subset U of a complex Banach space E into a complex Banach space F. Let $a \in U$. If $d_a f$ is a linear isomorphism between E and F, then f is a local analytic isomorphism at a.*

If E, F are real Banach spaces, then the same holds, if "analytic" is replaced by "real analytic".

Proof.   Replacing f by $(d_a f)^{-1} \cdot f$, it suffices to consider the case where $E = F$ and

$$d_a f = I,$$

the identity map.

By the continuity of df we can choose $r > 0$ such that, in the operator norm,

$$\|d_x f - I\| \leq \tfrac{1}{2} \qquad \text{on} \qquad \|x - a\| \leq r.$$

Then, by applying the mean value theorem to $f - id$,

$$(1) \qquad \|f(x_1) - f(x_2) - (x_1 - x_2)\| \leq \sup_{x \in \overline{x_1 x_2}} \|d_x f - I\| \, \|x_1 - x_2\|$$

$$\leq \tfrac{1}{2} \|x_1 - x_2\|$$

for $\|x_1 - a\| \leq r$ and $\|x_2 - a\| \leq r$. It follows that f is one-to-one on the ball $\|x - a\| \leq r$.

We now construct the inverse g of f on the ball $\|y - b\| \leq r/2$, where $b = f(a)$. To this end, we inductively define the maps

$$(2) \qquad g_k = g_{k-1} + id - f \circ g_{k-1}, \qquad k = 1, 2, \ldots,$$

where $g_0 \equiv a$, the constant map with value a.

Let us show by induction that all these maps are well defined on $\|y - b\| \leq r/2$ and satisfy

$$\||g_k - g_{k-1}\|| = \sup_{\|y - b\| \leq r/2} \|g_k(y) - g_{k-1}(y)\|$$

$$\leq \frac{r}{2^k}, \qquad k = 1, 2, \ldots .$$

Obviously, for $k = 1$, g_1 is well defined, and

$$\||g_1 - g_0\|| = \sup_{\|y - b\| \leq r/2} \|y - b\| = \frac{r}{2}.$$

For $k > 1$,

$$\||g_k - a\|| = \||g_k - g_0\|| \leq \sum_{j=1}^{k} \||g_j - g_{j-1}\||$$

$$\leq r \sum_{j \geq 1} \frac{1}{2^j} = r$$

by the triangle inequality and the induction hypotheses. So g_{k+1} is well defined on $\|y - b\| \leq r/2$, and

$$\||g_{k+1} - g_k\|| = \||f \circ g_k - f \circ g_{k-1} - (g_k - g_{k-1})\||$$

$$\leq \tfrac{1}{2}\||g_k - g_{k-1}\||$$

$$\leq \frac{r}{2^{k+1}}$$

using (1).

Thus, the g_k converge uniformly on $\|y - b\| \leq r/2$ to a continuous map g. Going to the limit in (2),

$$f \circ g = id.$$

It follows that f is a homeomorphism between an open neighborhood of a and the open ball $\|y - b\| < r/2$. Moreover, g is analytic on this ball by Theorem A.2, since the maps g_k are analytic. Consequently, f is a local analytic isomorphism at a.

It remains to show that g is real analytic when f is. Let $\tilde{f}$ be the analytic extension of f to some complex neighborhood of a. Then $d_a\tilde{f}$ is the identity on the complexification of E, hence $\tilde{f}$ is an analytic isomorphism on some complex neighborhood $\tilde{A}$ of a. On the other hand, f is also a homeomorphism on some real neighborhood A of a by the same arguments as before. Consequently, $f^{-1} = \tilde{f}^{-1}$ on $A \cap \tilde{A}$. Hence, f^{-1} is real analytic, and f is a real analytic isomorphism at a.   ∎

The implicit function theorem generalizes the inverse function theorem. Here, an analytic map

$$f: U \times V \to G$$

from the product of open subsets U, V of two Banach spaces E, F respectively into a Banach space G is considered. At a point (a, b) in $U \times V$ the partial derivative of f with respect to, say, the second coordinate is required to be a linear isomorphism between F and G. Here, the *partial derivative*

$$\partial_b f: F \to G$$

of f at (a, b) with respect to the second coordinate is obtained by fixing the first coordinate at a, so that f becomes an analytic map from V into G, whose derivative is taken at b.

Implicit Function Theorem.   *Let U, V be open subsets of complex Banach spaces E, F respectively, let*

$$f: U \times V \to G$$

be analytic, and let $(a, b) \in U \times V$. If the partial derivative of f at (a, b) with respect to b,

$$\partial_b f: F \to G,$$

is a linear isomorphism, then there exists an open neighborhood A of a and a unique continuous map $u: A \to V$ such that

$$f(x, u(x)) = f(a, b), \qquad u(a) = b$$

on A. Moreover, u is analytic on A.

If E, F are real Banach spaces, then the same holds, if "analytic" is replaced by "real analytic".

Proof.   We may replace f by $(\partial_b f)^{-1} \cdot f$. Then f maps $U \times V$ into F, and $\partial_b f$ is the identity map on F.

Now consider the map φ from $U \times V$ into $E \times F$ defined by

$$(x, y) \to (x, f(x, y)).$$

Its derivative

$$d_{a,b}\varphi = \begin{bmatrix} I_E & 0 \\ * & I_F \end{bmatrix}$$

at (a, b) is a linear isomorphism on E, F, so φ is a local (real) analytic isomorphism at (a, b) by the inverse function theorem. Its local inverse has the form

$$(x, z) \to (x, g(x, z)).$$

In particular,

$$f(x, g(x, z)) = z,$$

and for $c = f(a, b)$ we have

$$g(a, c) = b.$$

Hence, in a neighborhood A of a, we can define u by

$$u(x) = g(x, c).$$

The map u is (real) analytic on A, proving existence.

To prove uniqueness we first choose A to be connected and so small that φ is a local isomorphism at every point $(x, u(x))$ with x in A. Now suppose $v: A \to V$ is another map such that

$$f(x, v(x)) = f(a, b), \qquad v(a) = b$$

on A. Let B be the subset of A where u and v agree. B is closed by continuity and contains a. The set B is also open. For, if $x \in B$, then $u(x) = v(x)$ and consequently

$$\varphi(x, u(x)) = \varphi(x, v(x)).$$

Since φ is a homeomorphism in a neighborhood of $(x, u(x))$ by construction, u and v must agree also in a neighborhood of x. It follows that $B = A$, since A is connected. This proves uniqueness. ∎

Let U be an open subset of a real Banach space E. A *vectorfield on U* is simply a map

$$X: U \to E.$$

Geometrically, a vectorfield assigns a tangent, or "velocity" vector to each point in U.

A *solution curve* of the vectorfield X with *initial value a* in U is a differentiable map

$$\phi: J \to U$$

from an open interval J containing 0 into U such that

$$\phi(0) = a$$

and

$$\frac{d}{dt}\phi(t) = X(\phi(t)), \qquad t \in J.$$

Thus, the solution curve passes through a at time $t = 0$, and its velocity agrees with the vectorfield at every point it passes through.

A solution curve ϕ depends on the time t, the initial value a and the vectorfield X. We often use the notations $\phi^t(a)$ and $\phi^t(a, X)$ to make this dependence explicit.

The local existence and uniqueness of solution curves is guaranteed by a simple condition. A vectorfield X is *locally Lipschitz* on U, if for every point a in U there is a neighborhood A and a positive constant L such that

$$\|X(x) - X(y)\| \leq L\|x - y\|$$

for all x, y in A. Every continuously differentiable vectorfield is locally Lipschitz by the mean value theorem, and every locally Lipschitz vectorfield is continuous and locally bounded.

Local Existence and Uniqueness Theorem.   *Suppose X is a locally Lipschitz vectorfield on U. Then, for every $a \in U$, there exists an interval J containing 0 and a solution curve*

$$\phi: J \to U$$

of X with initial value a. This curve is unique on J.

If X is real analytic, then ϕ is a real analytic function of t and a.

We skip the standard proof, which may be found for example in [Di, La1]. As a matter of fact, we do not need the theorem at all, but include it for the sake of completeness only. In all cases of interest to us, the solution curves can be written down explicitly.

Every point on a solution curve may be considered as its initial value if the time is translated accordingly. It follows that if two solution curves pass through the same point, then they must be identical on their common interval of definition. Therefore, every solution curve has at most one extension on each side, and to each initial value a one can associate a unique maximal interval J_a on which the solution curve with initial value a is defined. J_a is the union of the domains of all possible solution curves through a. The interval J_a may be finite, infinite on one side or the whole real line.

In Chapters 4 and 6 we encounter all three possibilities. The solution curves of the vectorfields V_n are defined for all time, those of the vectorfield W_1 are defined on the half line $(-\infty, \mu_1)$, and those of W_n, $n > 1$, are defined on the finite intervals (μ_{n-1}, μ_{n+1}).

Another important consequence of this observation is the

Semi-Group-Property. *Let ϕ be a solution curve of a locally Lipschitz vectorfield X. Then, for every point a on it,*

$$\phi^{t+s}(a) = \phi^t(\phi^s(a))$$

for all s and t for which both sides are defined.

Proof. As functions of t, both sides are solution curves of the vectorfield X with initial value $\phi^s(a)$. By uniqueness, the two curves are equal. ∎

To shed some more light on the semi-group-property, suppose all solution curves of a vectorfield X on a domain D are defined for all time. Then, for every real t, we have a map

$$\Phi^t \colon D \to D,$$

where $\Phi^t(x)$ is obtained by following the solution curve with initial value x up to time t. Clearly,

$$\Phi^0 = id_D$$

by definition, and

$$\Phi^{t+s} = \Phi^t \circ \Phi^s$$

by the semi-group-property. Thus, we obtain a one-parameter-group of continuous[1] maps of D onto itself. In fact, each such map Φ^t is a homeomorphism of D with inverse Φ^{-t}, since

$$\Phi^t \circ \Phi^{-t} = \Phi^{-t} \circ \Phi^t = \Phi^0 = id_D.$$

Such a one-parameter-group Φ^t, $t \in \mathbb{R}$, of homeomorphisms of D is called a *flow on D*.

Conversely, every flow on D which is differentiable in t arises from a vectorfield on D. Simply set

$$X(x) = \frac{d}{dt}\,\Phi^t(x)\bigg|_{t=0}.$$

We leave the elementary proof to the reader.

---

[1] Continuity follows from the proof of the local existence and uniqueness theorem and the semi-group-property.

C Manifolds

In this appendix we present the basic facts about manifolds required to make our discussion of isospectral sets self contained. All the manifolds that are of interest to us arise as subsets of Hilbert spaces. For this reason, we shall only consider submanifolds of Banach spaces, and not develop the theory of manifolds in general.

To begin with, recall the following two terms. A *diffeomorphism* between two open subsets of Banach spaces is a continuously differentiable homeomorphism between these two sets, whose inverse is also continuously differentiable. A *splitting* of a Banach space is a direct sum decomposition into two subspaces, which are both closed.

By definition, a subset M of a Banach space E is a *submanifold of E*, if for every point x on M there is a *coordinate system* of the following kind. There is a diffeomorphism

$$\varphi \colon U \to V$$

between an open neighborhood U of x in E and an open subset V of another Banach space F and a splitting $F = F_h \oplus F_v$ such that

$$\varphi(U \cap M) = V \cap F_h.$$

In short, a submanifold locally looks like a "horizontal slice" of a Banach space.

Depending on the regularity of the coordinate systems, one obtains different classes of submanifolds. For example, a submanifold is *real analytic*, if for every point there is a real analytic coordinate system. All our manifolds are of that class.

Example 1. An open subset of a Banach space is a real analytic submanifold.

Example 2. A closed linear subspace is a real analytic submanifold if and only if it can be complemented. That is, one can find another closed linear subspace so that the two form a splitting. This is always possible in a Hilbert space, where one can take the orthogonal complement. This is not always possible in a Banach space. There are closed linear subspaces of Banach spaces, which can not be complemented [Ru].

Example 3. The sequence space S introduced in Chapter 3 is not given as a subset of a Banach space. However, there is a one-to-one correspondence between S and an open subset of $\mathbb{R} \times \ell^2$. This allows us to identify S with a real analytic submanifold.

Example 4. The isospectral set

$$M(p) = \{q \in L^2 : \mu(q) = \mu(p)\}$$

introduced in Chapter 3 is a real analytic submanifold of L^2. For, by Theorem 3.7, the translated map

$$q \to (\kappa \times \mu)(q) - (\kappa \times \mu)(p)$$

defines a *global*, real analytic coordinate system, which maps $M(p)$ into the horizontal subspace $\ell_1^2 \times \{0\}$ of $\ell_1^2 \times (\mathbb{R} \times \ell^2)$ (after identifying S with $\mathbb{R} \times \ell^2$).

Example 5. Let M be any compact connected submanifold of a Banach space E, and let φ be any coordinate system on M mapping into a Banach space $F = F_h \oplus F_v$. Then φ^{-1} is a homeomorphism between a closed ball in F_h and a closed subset of M, that is, the intersection of a closed subset of E with M. This intersection is compact by assumption, hence also the ball is compact. This implies that F_h is finite dimensional, since every locally compact Banach space is finite dimensional [Ru]. It follows that every compact connected submanifold is finite dimensional.[1] Conversely, an infinite dimensional compact subset of E can never be a submanifold.

To give an example, let $\ell^2_{\mathbb{C}}$ be the Hilbert space of all complex ℓ^2-sequences $z = (z_1, z_2, \ldots)$ and let r_n, $n \geq 1$, be a sequence of positive numbers such that

$$\sum_{n \geq 1} r_n^2 < \infty.$$

Consider the set

$$T = \{z \in \ell^2_{\mathbb{C}} : |z_n| = r_n, n \geq 1\}.$$

Topologically, T is the product of the infinitely many circles $|z_n| = r_n$, so T is an infinite dimensional torus. However, the radii of these circles tend to zero just fast enough to make T a compact subset of $\ell^2_{\mathbb{C}}$. So T is not a submanifold.

One can give other examples of this phenomenon. For instance, every finite intersection of the subsets

$$B_n = \left\{ x \in \ell^2 : x_{n-1}^2 + x_n^2 + x_{n+1}^2 = \frac{1}{2^n} \right\}, \qquad n \geq 2$$

of ℓ^2 is a real analytic submanifold of ℓ^2. Their infinite intersection, however, is compact, hence not a submanifold.

The notion of tangent space is of fundamental importance. Intuitively, the tangent space of a manifold M at a point x consists of the velocity vectors at x of all smooth curves on M which pass through x. The tangent space is a closed linear space which one thinks of being attached to the point x.

For submanifolds of a Banach space this notion is easy to make precise. Let M be such a submanifold, let x be a point on M, and let φ be a coordinate system around x mapping into a Banach space $F = F_h \oplus F_v$. The *tangent space of M at x* is the closed linear space

$$T_x M = (d_x \varphi)^{-1}(F_h).$$

---

[1] If M is a connected submanifold, then one can show that the spaces F_h for different coordinate systems are isomorphic. The dimension of M is then by definition the dimension of any of these spaces.

Equivalently,

$$T_x M = \{v \in E : d_x\varphi(v) \in F_h\}$$

$$= \ker d_x(\pi_v \cdot \varphi),$$

where π_v denotes the projection of F onto F_v along F_h.

We show that the definition of $T_x M$ is independent of the coordinate system chosen. Suppose ψ and φ are two coordinate systems around x mapping into $F = F_h \oplus F_v$ and $G = G_h \oplus G_v$ respectively. Then $\psi \circ \varphi^{-1}$ is a diffeomorphism between open subsets $V \subset F$ and $W \subset G$ such that

$$W \cap G_h = \psi \circ \varphi^{-1}(V \cap F_h).$$

It follows that

$$G_h \supset d_{\varphi(x)}(\psi \circ \varphi^{-1})(F_h)$$

$$= d_x\psi \cdot d_{\varphi(x)}\varphi^{-1}(F_h)$$

$$= d_x\psi \cdot (d_x\varphi)^{-1}(F_h),$$

or

$$(d_x\psi)^{-1}(G_h) \supset (d_x\varphi)^{-1}(F_h).$$

The opposite inclusion holds by symmetry, so we have equality. Hence the definition of $T_x M$ does not depend on the coordinate system.

In a Hilbert space we can also consider the orthogonal complement

$$N_x M = (T_x M)^{\perp}$$

of $T_x M$. This is the *normal space of M at x*.

Example 1. If U is an open subset of a Banach space E, then $T_x U = E$ for all x in U.

Example 2. If M is a closed linear subspace, then $T_x M = M$ for all x in M.

Example 3. For all σ in S, one has $T_\sigma S = \mathbb{R} \times \ell^2$. This is a particular instance of Example 1.

Example 4. The tangent and normal spaces of $M(p)$ are computed in the proof of Theorem 4.1*.

Manifolds often arise as the solution set of some nonlinear equation, typically as the level set, or "fiber", of a mapping between two Banach spaces. Not every such level set, however, is a submanifold, as the simple example

$$\{(x, y) : x^2 - y^2 = 0\}$$

of a subset in $\mathbb{R}^2$ shows. It is therefore useful to have a simple sufficient criterion for a level set to be a submanifold.

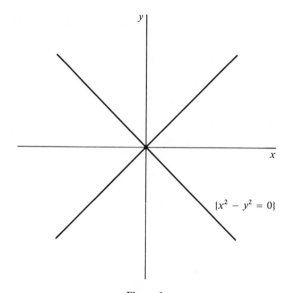

Figure 1.

Let $f: A \to F$ be a continuously differentiable map from an open subset A of a Banach space E into another Banach space F. A point c in F is a *regular value of f*, if for every point x in the level set

$$M_c = \{x \in A : f(x) = c\}$$

there exists a splitting $E = E_h \oplus E_v$, such that $d_x f \,|\, E_v$, the restriction of $d_x f$ to E_v, is a linear isomorphism between E_v and F.

Note that c is a regular value when M_c is empty.

If $F = \mathbb{R}$, then c is a regular value of f if and only if $d_x f \neq 0$ at every point x in M_c. More generally, if $F = \mathbb{R}^n$, then c is a regular value of f, if $d_x f$ has rank n (that is, is onto) at every point x in M_c.

Regular Value Theorem.   *Suppose $f: A \to F$ is a real analytic map from an open subset A of a Banach space E into another Banach space F. If $c \in F$ is a regular value of f, then*

$$M_c = \{x \in A : f(x) = c\}$$

is a real analytic submanifold of E. Moreover,

$$T_x M_c = \ker d_x f$$

at every point x in M_c.

Proof.   Without loss of generality we may assume that $c = 0$. By hypotheses, for every x in M_0 there is a splitting $E = E_h \oplus E_v$ such that with respect to this splitting the second partial derivative

$$L = \partial_{x_v} f : E_v \to F$$

is a linear isomorphism between E_v and F. We may therefore define a map φ from A into E by setting

$$\varphi(x_h, x_v) = (x_h, L^{-1} f(x_h, x_v)).$$

This map is real analytic. Moreover,

(*) $\varphi(M_0 \cap U) = \varphi(U) \cap E_h$

for every neighborhood U of x in A.

The derivative of φ at x is a linear isomorphism of E, since

$$d_x \varphi = \begin{pmatrix} I & 0 \\ * & I \end{pmatrix}.$$

By the inverse function theorem, φ is a real analytic isomorphism between an open neighborhood U of x and an open neighborhood V of $\varphi(x)$. Hence, in view of (*), φ is a real analytic coordinate system around x. This shows that M_0 is a real analytic submanifold of E.

Finally,

$$\begin{aligned}
T_x M_0 &= \ker d_x(\pi_v \circ \varphi) \\
&= \ker \pi_v \cdot d_x \varphi \\
&= \ker L^{-1} d_x f \\
&= \ker d_x f,
\end{aligned}$$

since L^{-1} is a linear isomorphism. Here, π_v denotes the projection of E onto E_v.   ■

Example 4. The isospectral set $M(p)$ is the fiber of the real analytic map $\mu: L^2 \to S$ over $\mu(p)$. The proof of Theorem 4.1 shows that every point in S is a regular value of μ. Hence, $M(p)$ is a real analytic submanifold of L^2.

Example 5. On ℓ_C^2, the functions

$$f_n(z) = |z_n|^2, \qquad n \geq 1,$$

are *real analytic*, since $|z_n|^2 = x_n^2 + y_n^2$ is a polynomial in the real and imaginary part of $z_n = x_n + iy_n$. Their gradients are given by

$$\frac{\partial f_n}{\partial z}(\zeta) = 2\,\mathrm{Re}\,\bar{z}_n\zeta_n, \qquad \zeta \in \ell_C^2.$$

They do not vanish, in fact, they are all linearly independent at every point z in ℓ_C^2 all of whose components do not vanish.

It follows that every hypersurface

$$T_n = \{z \in \ell_C^2 : f_n(z) = r_n^2\}$$

$$= \{z \in \ell_C^2 : |z_n| = r_n\}, \qquad r_n > 0$$

is a real analytic submanifold of ℓ_C^2. The same is true for every finite intersection

$$\bigcap_{1 \leq n \leq N} T_n = \{z \in \ell_C^2 : |z_n| = r_n, 1 \leq n \leq N\}$$

of them. However, their infinite intersection

$$T = \bigcap_{n \geq 1} T_n$$

is *not* a submanifold of ℓ_C^2, if $\sum r_n^2 < \infty$, since this set is compact.

This example shows that the linear independence of gradients is not enough to guarantee that a level set is a submanifold.

Example 6. For p in L^2 and $n \geq 1$,

$$M_n(p) = \{q \in L^2 : \mu_n(q) = \mu_n(p)\}$$

is a level set of the real analytic function μ_n on L^2. Its derivative

$$d_q\mu_n(v) = \langle g_n^2(x, q), v \rangle$$

is never identically zero. Thus, every real number is a regular value of μ_n, and $M_n(p)$ is a real analytic submanifold of L^2 for every p. Moreover,

$$T_q M_n(p) = \ker d_q \mu_n$$

$$= \{v \in L^2 : \langle g_n^2(q), v \rangle = 0\}.$$

Clearly, $N_q M_n(p)$ is the line spanned by $g_n^2(q)$.

A *vectorfield* on a submanifold M of a Banach space E is a map

$$X : M \to E$$

such that

$$X(x) \in T_x M$$

for every $x \in M$. A *solution curve* of X with *initial value* a on M is a differentiable map

$$\phi : J \to M$$

from an open interval J containing 0 into M such that $\phi(0) = a$ and

$$\frac{d}{dt} \phi(t) = X(\phi(t))$$

for all $t \in J$.

The local existence and uniqueness theorem of Appendix B generalizes immediately to submanifolds. Here, a vectorfield is *locally Lipschitz*, if for every point a on the submanifold M there is a neighborhood A of a in E and a positive constant L such that

$$\|X(x) - X(y)\| \le L\|x - y\|$$

for all $x, y \in A \cap M$.

Local Existence and Uniqueness Theorem.   *Suppose X is a locally Lipschitz vectorfield on a submanifold M of a Banach space E. Then, for every $a \in M$ there exists a solution curve*

$$\phi : J \to M$$

of X with initial value a. This curve is unique on J.

If M and X are real analytic, then ϕ is a real analytic function of t and a.

The vectorfield X is real analytic, if there exists an open neighborhood U of M and a real analytic map $\tilde{X} : U \to E$ such that $\tilde{X} | M = X$.

We omit the standard proof. In short, it consists of three simple steps. Using a local coordinate system, the vectorfield X is pushed forward to an open subset of a Banach space. The local existence and uniqueness theorem of Appendix B is applied to obtain its solution curves, which are then pulled back to the submanifold M.

As a matter of fact, in Chapters 4 and 6 we proceed in just this way instead of applying the general theorem, because in both cases the situation is particularly simple. We have a *global* coordinate system which maps the given (complicated looking) vectorfield into a *constant* vectorfield on a Banach space. The unique solution curves of the latter are straight lines, which are pulled back to the submanifold. This approach in addition yields the analytic dependence of these curves on time, initial value *and* vectorfield.

D Some Functional Analysis

In this appendix we collect some facts about compact operators and state the Fredholm Alternative in a form, which is useful for us.

Let H be a Hilbert space. A linear operator T on H is *compact*, if it maps weakly converging sequences into strongly converging sequences. Equivalently, a compact operator T maps bounded subsets into relatively compact subsets of H.[1]

Obviously, compact operators are bounded.

An operator has *finite rank*, if its range is finite dimensional. Clearly, bounded operators with finite rank operators are compact. From these, more interesting compact operators are obtained by

Lemma 1. *The uniform limit of compact operators is compact.*

Proof. Let T be the uniform limit of compact operators T_n. Let x_m be a weakly converging sequence, and let x be its weak limit. Then, by the

---

[1] The latter is the usual definition of compactness. The equivalence to the former requires a short proof, which we forego, since it is standard.

principle of uniform boundedness,

$$\|x\| \leq \sup_m \|x_m\| \leq M < \infty,$$

hence

$$\|Tx_m - Tx\| \leq \|Tx_m - T_n x_n\| + \|T_n x_m - T_n x\| + \|T_n x - Tx\|$$

$$\leq \|T - T_n\| \cdot M + \|T_n x_m - T_n x\| + \|T_n - T\| \cdot M.$$

The first and third term can be made small by choosing n sufficiently large. Then the middle term can be made equally small by choosing m sufficiently large, since all T_n are compact. It follows that

$$\|Tx_m - Tx\| \to 0$$

as m tends to infinity. That is, Tx_n converges strongly to Tx. Thus, T is compact. ∎

Suppose H is a separable Hilbert space (as are all Hilbert spaces we encounter in this book). A linear operator T on H is *Hilbert-Schmidt* if

$$\sum_{n \geq 1} \|Te_n\|^2 < \infty$$

for some orthonormal basis e_n, $n \geq 1$, of H.

A Hilbert-Schmidt operator is bounded. To see this, write

$$x = \sum_{n \geq 1} x_n e_n.$$

Then $\|x\|^2 = \sum_{n \geq 1} |x_n|^2$, and by the triangle and Schwarz inequality,

$$\|Tx\| = \left\| \sum_{n \geq 1} x_n Te_n \right\|$$

$$\leq \sum_{n \geq 1} |x_n| \|Te_n\|$$

$$\leq \left(\sum_{n \geq 1} |x_n|^2 \right)^{1/2} \left(\sum_{n \geq 1} \|Te_n\|^2 \right)^{1/2}$$

$$\leq C\|x\|.$$

Hence, T is bounded.

Furthermore, we have

Theorem 1. *If T is Hilbert-Schmidt, then T is compact.*

Proof. Let $e_1, e_2, \ldots$ be an orthonormal basis of H such that $\sum \|Te_m\|^2$ is finite. For $n \geq 1$, let T_n be the linear operator defined by

$$T_n e_m = \begin{cases} Te_m, & m \leq n \\ 0, & m > n \end{cases}.$$

Each T_n is bounded and has finite rank, so is compact. Moreover, using the Schwarz inequality,[2]

$$\begin{aligned} \|T_n - T\|^2 &= \|T_n^* - T^*\|^2 \\ &= \sup_{\|x\| \leq 1} \|T_n^* x - T^* x\|^2 \\ &= \sup_{\|x\| \leq 1} \sum_{m \geq 1} |\langle T_n^* x - T^* x, e_m \rangle|^2 \\ &= \sup_{\|x\| \leq 1} \sum_{m \geq 1} |\langle x, T_n e_m - Te_m \rangle|^2 \\ &\leq \sum_{m \geq 1} \|T_n e_m - Te_m\|^2 \\ &= \sum_{m > n} \|Te_m\|^2 \to 0 \end{aligned}$$

as n tends to infinity. Thus, T is the uniform limit of compact operators, hence compact by Lemma 1. ∎

We frequently have to consider compact perturbations of the identity map. These are operators of the form $I - T$, where I is the identity and T is compact. By slight abuse of terminology, we call them *Fredholm operators*.[3] In some respect, they resemble operators on finite dimensional spaces:

Theorem 2 (Fredholm Alternative). *Suppose A is a Fredholm operator. Then the following three statements are equivalent.*

(1) *A is boundedly invertible.*
(2) *A is onto.*
(3) *A is one-to-one.*

---

[2] As usual, $\|T\| = \sup_{\|x\| = 1} \|Tx\|$ denotes the operator norm of T.

[3] Usually, this term is reserved for the much larger class of operators with finite index and coindex. See for instance [Ka].

Proof. (1) $\Rightarrow$ (2)   Obvious.

(2) $\Rightarrow$ (3)   Suppose $A = I - T$ is onto. The kernels of A^n, $n = 0, 1, \ldots,$ form an increasing sequence of closed subspaces of H. Suppose no two of them are equal. Then we can choose unit vectors

$$x_n \in \ker(A^n) \cap \ker(A^{n-1})^\perp, \qquad n \geq 1.$$

We have $x_n \to 0$ weakly, hence $Tx_n \to 0$ strongly by compactness. Then also

$$\langle Tx_n - Tx_{n-1}, x_n \rangle \to 0.$$

On the other hand, A maps $\ker(A^n)$ into $\ker(A^{n-1})$, and $T = I - A$ maps $\ker(A^n)$ into itself. Therefore,

$$\langle Tx_n - Tx_{n-1}, x_n \rangle = \langle Tx_n, x_n \rangle = \langle x_n, x_n \rangle = 1,$$

a contradiction. Thus, we have

$$\ker(A^n) = \ker(A^{n-1})$$

for some $n \geq 1$. That is, $AA^{n-1}x = 0$ implies $A^{n-1}x = 0$ for all x. Since A^{n-1} is onto for every $n \geq 1$, it follows that A is one-to-one.

(3) $\Rightarrow$ (1)   Suppose $A = I - T$ is one-to-one. Suppose we have

$$\inf_{\|x\| = 1} \|Ax\| = 0.$$

Then we can choose a sequence of unit vectors x_n such that $Ax_n \to 0$ strongly. Extracting a subsequence, Tx_n converges to some x by compactness. Then also

$$x_n = Ax_n + Tx_n \to x$$

strongly, hence $\|x\| = 1$. On the other hand,

$$Ax = \lim Ax_n = 0,$$

hence $x = 0$, since A is one-to-one. This is a contradiction, and we must have

$$\inf_{\|x\| = 1} \|Ax\| \geq \delta > 0.$$

It follows that the pointwise inverse of A is bounded on its range $rg(A)$, for

$$\|A^{-1}Ax\| = \|x\| \leq \delta^{-1}\|Ax\|$$

for all x. This further implies that $rg(A)$ is closed.

It remains to show that $rg(A)$ is all of H. Consider $rg(A^n)$ for $n = 0, 1, \ldots$. This is a decreasing sequence of *closed* subspaces of H, since T^n is compact, hence A^n is Fredholm for all $n \geq 1$. As before, one shows that

$$rg(A^n) = rg(A^{n-1})$$

must hold for some $n \geq 1$. Hence, for each y there is an x such that $A^n y = A^{n-1} x$. Since A^n is one-to-one for every $n \geq 1$, we have $y = Ax$. This shows that A is onto.   ∎

Among other things, the Fredholm Alternative can be used to show that a sequence of vectors is a basis. The following theorem describes the general situation.

Theorem 3.   *Let e_n, $n \geq 1$, be an orthonormal basis of a Hilbert space H. Suppose d_n, $n \geq 1$, is another sequence of vectors in H that either spans or is linearly independent. If, in addition,*

$$\sum_{n \geq 1} \|d_n - e_n\|^2 < \infty,$$

then d_n, $n \geq 1$, is also a basis of H. Moreover, the map

$$x \rightarrow (\langle x, d_n \rangle, n \geq 1)$$

is a linear isomorphism between H and ℓ^2.

Proof.   Define an operator A on H by

$$Ax = \sum_{n \geq 1} \langle x, e_n \rangle \, d_n.$$

A maps e_n into d_n. This operator is a compact perturbation of the identity, for

$$\sum_{n \geq 1} \|(A - I)e_n\|^2 = \sum_{n \geq 1} \|d_n - e_n\|^2 < \infty,$$

so $A - I$ is Hilbert-Schmidt and hence compact by Theorem 1.

If the sequence d_n is linearly independent, then A is also one-to-one: if $Ax = 0$, then

$$\langle x, e_n \rangle = 0, \qquad n \geq 1$$

by linear independence, and consequently $x = 0$, since the e_n are a basis. If, on the other hand, the sequence d_n spans, then the range of A is dense in H.

It follows that A is onto, since the range of a compact perturbation of the identity is always closed [Ru].[4]

So in both cases the operator A is boundedly invertible by the Fredholm alternative. Hence the d_n are also a basis.

Finally, we have

$$\langle x, d_n \rangle = \langle x, Ae_n \rangle = \langle A^*x, e_n \rangle.$$

The map

$$x \rightarrow (\langle A^*x, e_n \rangle, n \geq 1)$$

is a linear isomorphism between H and ℓ^2, since A^* is boundedly invertible, and the e_n are an orthonormal basis.   ■

---

[4] This is not proven here, because this part of the theorem is used only once in the proof of Lemma 4.3.

E Three Lemmas on Infinite Products

Lemma 1. (a) *Suppose* a_{mn}, $m, n > 1$, *are complex numbers satisfying*

$$|a_{mn}| = O\left(\frac{1}{|m^2 - n^2|}\right), \qquad m \neq n.$$

Then

$$\prod_{\substack{m \geq 1 \\ m \neq n}} (1 + a_{mn}) = 1 + O\left(\frac{\log n}{n}\right), \qquad n \geq 1.$$

(b) *In addition, if* b_n, $n \geq 1$, *is a square summable sequence of complex numbers, then*

$$\prod_{\substack{m, n \geq 1 \\ m \neq n}} (1 + a_{mn} b_n) < \infty.$$

Proof. (a) By assumption,

$$\sum_{\substack{m \geq 1 \\ m \neq n}} |a_{mn}| \leq C \sum_{\substack{m \geq 1 \\ m \neq n}} \frac{1}{|m^2 - n^2|}$$

with some positive constant C. The sum on the right can be estimated by

$$\sum_{\substack{m \geq 1 \\ m \neq n}} \frac{1}{|m^2 - n^2|} = \sum_{\substack{1 \leq m \leq 2n \\ m \neq n}} \frac{1}{|m - n|} \frac{1}{m + n} + \sum_{m > 2n} \frac{1}{m^2 - n^2}$$

$$\leq \frac{2}{n} \sum_{1 \leq k \leq n} \frac{1}{k} + \sum_{k > n} \frac{1}{k^2}$$

$$\leq \frac{2}{n} (1 + \log n) + \frac{1}{n}.$$

Hence we obtain, with a different constant C',

$$\left| \prod_{\substack{m \geq 1 \\ m \neq n}} (1 + a_{mn}) - 1 \right| \leq \prod (1 + |a_{mn}|) - 1$$

$$\leq \exp\left(\sum |a_{mn}| \right) - 1$$

$$\leq \exp\left(C' \frac{1 + \log n}{n} \right) - 1$$

$$= O\left(\frac{\log n}{n} \right).$$

(b)   By the proof of (a), $\sum_{1 \leq m \neq n} |a_{mn}| \leq C(1 + \log n)/n$. Hence, by the Schwarz inequality,

$$\sum_{\substack{m, n \geq 1 \\ m \neq n}} |a_{mn}||b_n| = \sum_{n \geq 1} |b_n| \left(\sum_{\substack{m \geq 1 \\ m \neq n}} |a_{mn}| \right)$$

$$\leq C \sum_{n \geq 1} \frac{1 + \log n}{n} |b_n|$$

$$\leq C \left(\sum_{n \geq 1} \left(\frac{1 + \log n}{n} \right)^2 \sum_{n \geq 1} |b_n|^2 \right)^{1/2}$$

$$< \infty.$$

The claim follows.   ∎

Problem 1.   Show that the error term $O(\log n/n)$ in Lemma 1 can not be improved.

Recall the well known product expansion [Ah, Ti]

$$\frac{\sin \sqrt{\lambda}}{\sqrt{\lambda}} = \prod_{m \geq 1} \frac{m^2 \pi^2 - \lambda}{m^2 \pi^2}.$$

If in the numerator the numbers $m^2 \pi^2$ are replaced by complex numbers z_m with the same asymptotic behavior, then we obtain an entire function, which approximates $\sin \sqrt{\lambda}/\sqrt{\lambda}$ on certain circles. This is the content of the following lemma.

Lemma 2. *Suppose z_m, $m \geq 1$, is a sequence of complex numbers such that*

$$z_m = m^2 \pi^2 + O(1).$$

Then the infinite product

$$\prod_{m \geq 1} \frac{z_m - \lambda}{m^2 \pi^2}$$

is an entire function of λ, whose roots are precisely the z_m, $m \geq 1$. Moreover,

$$\prod_{m \geq 1} \frac{z_m - \lambda}{m^2 \pi^2} = \frac{\sin \sqrt{\lambda}}{\sqrt{\lambda}} \left(1 + O\left(\frac{\log n}{n}\right)\right)$$

uniformly on the circles $|\lambda| = (n + \frac{1}{2})^2 \pi^2$.

Proof. By the uniform boundedness of $z_m - m^2 \pi^2$ for $m \geq 1$,

$$\sum_{m \geq 1} \left| \frac{z_m - \lambda}{m^2 \pi^2} - 1 \right| = \sum_{m \geq 1} \left| \frac{z_m - m^2 \pi^2 - \lambda}{m^2 \pi^2} \right|$$

converges uniformly on bounded subsets of $\mathbb{C}$. Therefore, the infinite product converges to an entire function of λ, whose roots are precisely z_m, $m \geq 1$.

The quotient of the given product and $\sin \sqrt{\lambda}/\sqrt{\lambda}$ is the infinite product

$$\prod_{m \geq 1} \frac{z_m - \lambda}{m^2 \pi^2 - \lambda}.$$

On the circles $|\lambda| = (n + \frac{1}{2})^2 \pi^2$ the uniform estimates

$$\frac{z_m - \lambda}{m^2 \pi^2 - \lambda} = \begin{cases} 1 + O\left(\dfrac{1}{n}\right), & m = n \\[2ex] 1 + O\left(\dfrac{1}{|m^2 - n^2|}\right), & m \neq n \end{cases}$$

hold. Then, by Lemma 1,

$$\prod_{m \geq 1} \frac{z_m - \lambda}{m^2 \pi^2 - \lambda} = \left(1 + O\!\left(\frac{1}{n}\right)\right)\!\left(1 + O\!\left(\frac{\log n}{n}\right)\right)$$

$$= 1 + O\!\left(\frac{\log n}{n}\right)$$

uniformly on these circles. ∎

The last lemma is a variant of the preceding one.

Lemma 3. *Suppose z_m, $m \geq 1$, is a sequence of complex numbers such that*

$$z_m = m^2 \pi^2 + O(1).$$

Then, for each $n \geq 1$

$$\prod_{\substack{m \geq 1 \\ m \neq n}} \frac{z_m - \lambda}{m^2 \pi^2}$$

is an entire function of λ such that

$$\prod_{\substack{m \geq 1 \\ m \neq n}} \frac{z_m - \lambda}{m^2 \pi^2} = \tfrac{1}{2}(-1)^{n+1}\left(1 + O\!\left(\frac{\log n}{n}\right)\right)$$

uniformly for $\lambda = n^2 \pi^2 + O(1)$.

Proof. We only prove the last statement. By the product expansion for $\sin \sqrt{\lambda}/\sqrt{\lambda}$, we have

$$\prod_{\substack{m \geq 1 \\ m \neq n}} \frac{m^2 \pi^2 - n^2 \pi^2}{m^2 \pi^2} = -n^2 \pi^2 \frac{d}{d\lambda} \frac{\sin \sqrt{\lambda}}{\sqrt{\lambda}}\bigg|_{\lambda = n^2 \pi^2} = \tfrac{1}{2}(-1)^{n+1}.$$

The quotient of this and the given product is

$$\prod_{\substack{m \geq 1 \\ m \neq n}} \frac{z_m - \lambda}{m^2 \pi^2 - n^2 \pi^2} = 1 + O\!\left(\frac{\log n}{n}\right)$$

uniformly for $\lambda = n^2 \pi^2 + O(1)$ by Lemma 1. ∎

F Gaussian Elimination

In Chapters 5 and 6 explicit formulas for the isomorphisms $\exp_q$ and μ_E^{-1} are given, which only involve data at a single point q. In both cases the crucial ingredient of their derivation is an identity relating the determinant of an $n \times n$-matrix to an n-fold product.

In this appendix proofs of these identities are given. They are based on the familiar Gaussian elimination scheme. For our purposes it can be stated as follows.

Lemma 1 (The Gaussian Elimination Scheme). *Let $A = (a_{ij}^0)$ be an $n \times n$-matrix. For $1 \leq k < n$, let*

(1)
$$a_{ij}^k = a_{ij}^{k-1} - \frac{a_{ik}^{k-1} a_{kj}^{k-1}}{a_{kk}^{k-1}}, \qquad k < i, j \leq n,$$

assuming that $a_{kk}^{k-1} \neq 0$ for $1 \leq k < n$. Then A can be transformed into the upper triangular matrix

$$(2) \qquad \begin{bmatrix} a_{11}^0 & a_{12}^0 & a_{13}^0 & \cdots & a_{1n}^0 \\ 0 & a_{22}^1 & a_{23}^1 & \cdots & a_{2n}^1 \\ 0 & 0 & a_{33}^2 & \cdots & a_{3n}^2 \\ \vdots & \vdots & \vdots & \ddots & \vdots \\ 0 & 0 & 0 & \cdots & a_{nn}^{n-1} \end{bmatrix}$$

by elementary row transformations.[1] Consequently,

$$\prod_{k=1}^{n} a_{kk}^{k-1} = \det A.$$

More generally, if

$$(1^*) \qquad a_{ij}^k = \alpha_i^k \left(a_{ij}^{k-1} - \frac{a_{ik}^{k-1} a_{kj}^{k-1}}{a_{kk}^{k-1}} \right) \beta_j^k, \qquad k < i, j \le n$$

with real or complex numbers α_i^k, β_j^k, then

$$\prod_{k=1}^{n} a_{kk}^{k-1} = \det A \prod_{1 \le i < j \le n} \alpha_j^i \beta_j^i.$$

Proof. Apply induction on n. The statement is true for 1×1-matrices. So assume it is true for $(n-1) \times (n-1)$-matrices, $n > 1$.

By assumption, $a_{11}^0 \ne 0$. In case (1) holds, multiplying the first row of A by $-(a_{i1}/a_{11}^0)$ and adding it to the ith row, $1 < i \le n$, we obtain the matrix

$$(3) \qquad \begin{bmatrix} a_{11}^0 & a_{12}^0 & a_{13}^0 & \cdots & a_{1n}^0 \\ 0 & a_{22}^1 & a_{23}^1 & \cdots & a_{2n}^1 \\ 0 & a_{32}^1 & a_{33}^1 & \cdots & a_{3n}^1 \\ \vdots & \vdots & \vdots & \ddots & \vdots \\ 0 & a_{n2}^1 & a_{n3}^1 & \cdots & a_{nn}^1 \end{bmatrix}.$$

To the lower right $(n-1) \times (n-1)$-matrix the induction hypotheses applies. Hence, A can be transformed into the matrix (2) by elementary row transformations. These transformations do not affect the determinant of a matrix, so

$$\prod_{k=1}^{n} a_{kk}^{k-1} = \det A.$$

---

[1] An elementary row transformation consists in adding a multiple of one row to another row.

In case (1*) holds, first multiply the matrix A from left and right with the diagonal matrices

$$\begin{bmatrix} 1 & & & & \\ & \alpha_2^1 & & & \\ & & \alpha_3^1 & & \\ & & & \ddots & \\ & & & & \alpha_n^1 \end{bmatrix}, \quad \begin{bmatrix} 1 & & & & \\ & \beta_2^1 & & & \\ & & \beta_3^1 & & \\ & & & \ddots & \\ & & & & \beta_n^1 \end{bmatrix}$$

respectively. You obtain a matrix $\tilde{A} = (\tilde{a}_{ij})$ such that

$$a_{ij}^1 = \tilde{a}_{ij} - \frac{\tilde{a}_{i1}\tilde{a}_{1j}}{\tilde{a}_{11}}, \quad 1 < i, j \leq n.$$

By the same row transformations as before, $\tilde{A}$ is transformed into the matrix (3). Hence,

$$\det A \prod_{1 < j \leq n} \alpha_j^1 \beta_j^1 = a_{11}^0 \det A^1,$$

where A^1 denotes the lower right $(n - 1) \times (n - 1)$-matrix in (3). To A^1 the induction hypotheses applies, and we have

$$\det A^1 \prod_{2 \leq i < j \leq n} \alpha_j^i \beta_j^i = \prod_{2 \leq k \leq n} a_{kk}^{k-1}.$$

The claim follows. ■

We apply the Gaussian elimination scheme to the matrix

$$\Theta^{(n)} = (\theta_{ij})_{1 \leq i,j \leq n}$$

with elements

$$\theta_{ij} = \theta_{ij}(x, \xi, q)$$

$$= \delta_{ij} + (e^{\xi_i} - 1) \int_x^1 g_i(s, q) g_j(s, q) \, ds.$$

To this end we verify an identity of the form (1) relating the θ-functions of

$$q_0 = q, \quad q_k = \phi^{\xi_k}(q_{k-1}, V_k), \quad 1 \leq k \leq n,$$

as they were defined in the proof of Theorem 5.2.

Lemma 2. *For $1 \leq k \leq n$ and $k < i, j \leq n$,*

$$\theta_{ij}(q_k) = \theta_{ij}(q_{k-1}) - \frac{\theta_{ik}(q_{k-1})\theta_{kj}(q_{k-1})}{\theta_{kk}(q_{k-1})},$$

where $q_k = \phi^{\xi_k}(q_{k-1}, V_k)$.[2]

We drop the arguments x, ξ, which are fixed in the following discussion. Combining Lemma 2 with the Gaussian elimination scheme we obtain

$$\det \Theta^{(n)} = \prod_{k=1}^{n} \theta_{kk}(q_{k-1})$$

$$= \prod_{k=1}^{n} \theta_k(x, \xi_k, q_{k-1})$$

in the notation of chapter 5 as we wanted to show.

Proof of Lemma 2. By Theorem 5.1,

$$g_i(q_k) = g_i - (e^{\xi_k} - 1)\frac{g_k}{\theta_k}\int_x^1 g_i g_k \, ds$$

$$= g_i - \frac{g_k \theta_{ki}}{\theta_{kk}}$$

for $i \neq k$, where the functions on the right hand side are all evaluated at q_{k-1}. Hence, for $k < i, j \leq n$, we have

$$\theta_{ij}(q_k) = \delta_{ij} + (e^{\xi_i} - 1)\int_x^1 g_i(q_k)g_j(q_k)\, ds$$

$$= \delta_{ij} + (e^{\xi_i} - 1)\int_x^1 \left(g_i - \frac{g_k\theta_{ki}}{\theta_{kk}}\right)\left(g_j - \frac{g_k\theta_{kj}}{\theta_{kk}}\right) ds$$

$$= \theta_{ij} - \int_x^1 (e^{\xi_i} - 1)\cdot\left(g_i g_k\frac{\theta_{kj}}{\theta_{kk}} + g_j g_k\frac{\theta_{ki}}{\theta_{kk}} - g_k^2\frac{\theta_{ki}\theta_{kj}}{\theta_{kk}^2}\right) ds.$$

Using the identities

$$(e^{\xi_i} - 1)\theta_{ki} = (e^{\xi_k} - 1)\theta_{ik}, \qquad i \neq k$$

---

[2] We thank T. Nanda for his help in proving this lemma.

and

$$\theta'_{kl} = -(e^{\xi k} - 1)g_k g_l,$$

the second term becomes

$$\int_x^1 \left((e^{\xi i} - 1)g_i g_k \frac{\theta_{kj}}{\theta_{kk}} + (e^{\xi k} - 1)g_k g_j \frac{\theta_{ik}}{\theta_{kk}} - (e^{\xi k} - 1)g_k^2 \frac{\theta_{ik}\theta_{kj}}{\theta_{kk}^2} \right) ds$$

$$= \int_x^1 \frac{d}{ds} \left(-\frac{\theta_{ik}\theta_{kj}}{\theta_{kk}} \right) ds$$

$$= \frac{\theta_{ik}\theta_{kj}}{\theta_{kk}} \bigg|_x.$$

All the functions are evaluated at q_{k-1}, so the lemma is proven. ∎

Next we apply Lemma 1 to the matrix

$$\Omega^{(n)} = (\omega_{ij})_{1 \le i, j \le n},$$

where

$$\omega_{ij} = \omega_{ij}(x, \sigma, p)$$

$$= \frac{\sigma_i - \mu_i}{\sigma_i - \mu_j} [w_i, z_j] \bigg|_{\lambda = \sigma_i}.$$

See Chapter 6 for the definition of w_i and z_j. We shall verify an identity of the form (1*) relating the ω-functions of

$$p_0 = p, \qquad p_k = \phi^{\sigma_k - \mu_k}(p_{k-1}, W_k), \qquad 1 \le k \le n$$

to each other.

Recall that in general the flows of the vectorfields $W_1, \ldots, W_n$ are applied in some permuted order instead of their natural order to avoid the crossing of eigenvalues. As a consequence the Gaussian elimination scheme has to be applied to the columns of $\Omega^{(n)}$ in the same permuted order. Equivalently, the columns of $\Omega^{(n)}$ first have to be permuted properly so as to apply the scheme from left to right as usual. This permutation, however, affects only the sign of the determinant of $\Omega^{(n)}$, which is of no importance to us. Thus, for the sake of simplicity, we may assume that the flows of $W_1, \ldots, W_n$ are applied in their natural order.

Lemma 3. *For* $1 \leq k \leq n$ *and* $k < i, j \leq n$,

$$\omega_{ij}(p_k) = \frac{\mu_k - \sigma_i}{\sigma_k - \sigma_i} \frac{\sigma_k - \mu_j}{\mu_k - \mu_j} \left(\omega_{ij}(p_{k-1}) - \frac{\omega_{ik}(p_{k-1})\omega_{kj}(p_{k-1})}{\omega_{kk}(p_{k-1})} \right),$$

where $p_k = \phi^{\sigma_k - \mu_k}(p_{k-1}, W_k)$.[3]

Again, we dropped the arguments x, σ, which are fixed in the following. Here, the more general form of Lemma 1 applies, with

$$\alpha_i^k = \frac{\mu_k - \sigma_i}{\sigma_k - \sigma_i}, \qquad \beta_j^k = \frac{\sigma_k - \mu_j}{\mu_k - \mu_j}.$$

It follows that

$$\prod_{k=1}^{n} \omega_{kk}(p_k) = \det \Omega^{(n)} \prod_{1 \leq i < j \leq n} \frac{\mu_i - \sigma_j}{\sigma_i - \sigma_j} \frac{\sigma_i - \mu_j}{\mu_i - \mu_j}$$

as we wanted to show.

Proof of Lemma 3. By definition,

$$\omega_{ij}(p_k) = \frac{\sigma_i - \mu_i}{\sigma_i - \mu_j} [w_i(p_k), z_j(p_k)],$$

where here and in the following, the functions w_i are evaluated at $\lambda = \sigma_i$ without further mention.

By the remark following Theorem 6.2,

$$(4) \qquad z_j(p_k) = z_j - \frac{\sigma_k - \mu_k}{\mu_j - \mu_k} \frac{w_k}{\omega_{kk}} [z_j, z_k], \qquad j \neq k,$$

where the functions on the right hand side are evaluated at p_{k-1}. By the same arguments,

$$v_i(p_k) = w_i - \frac{\sigma_k - \mu_k}{\sigma_i - \mu_k} \frac{w_k}{\omega_{kk}} [w_i, z_k], \qquad i \neq k$$

is a genuine solution of $-y'' + p_k y = \lambda y$ for $\lambda = \sigma_i$. We claim that

$$(5) \qquad w_i(p_k) = \frac{\mu_k - \sigma_i}{\sigma_k - \sigma_i} v_i(p_k), \qquad i \neq k.$$

---

[3] This lemma is an unpublished result of J. Ralston and E. Trubowitz.

To prove this it suffices to check the boundary values of v_i. One calculates

$$v_i\bigg|_{x=0} = 1 - \frac{\sigma_k - \mu_k}{\sigma_i - \mu_k} = \frac{\sigma_i - \sigma_k}{\sigma_i - \mu_k}$$

and

$$v_i\bigg|_{x=1} = y_1(1, \mu_i) - \frac{\sigma_k - \mu_k}{\sigma_i - \mu_k} y_1(1, \mu_k) y_1(1, \mu_i) y_2'(1, \mu_k)$$

$$= \left(1 - \frac{\sigma_k - \mu_k}{\sigma_i - \mu_k}\right) y_1(1, \mu_i)$$

$$= \frac{\sigma_i - \sigma_k}{\sigma_i - \mu_k} y_1(1, \mu_i, p_{k-1}),$$

using the Wronskian identity and the fact that the boundary values of ω_{kk} are always 1. By equation (4),

$$y_2'(1, \mu_i, p_{k-1}) = y_2'(1, \mu_i, p_k), \qquad i \neq k$$

and therefore also

$$y_1(1, \mu_i, p_{k-1}) = y_1(1, \mu_i, p_k)$$

again by the Wronskian identity. It follows that the right hand side in (5) has the same values at 0 and 1 as $w_i(p_k)$, namely 1 and $y_1(1, \mu_i, p_k)$ respectively. This proves (5).

We determine the Wronskian of $v_i(p_k)$ and $z_j(p_k)$. An elementary but lengthy calculation shows that

$$[v_i(p_k), z_j(p_k)] = \left[w_i - \frac{\sigma_k - \mu_k}{\sigma_i - \mu_k} \frac{w_k}{\omega_{kk}} [w_i, z_k], z_j - \frac{\sigma_k - \mu_k}{\mu_j - \mu_k} \frac{w_k}{\omega_{kk}} [z_j z_k] \right]$$

$$= [w_i, z_j] - \frac{\sigma_k - \mu_k}{\sigma_i - \mu_k} \frac{[w_i, z_k][w_k, z_j]}{\omega_{kk}}$$

$$- \frac{\sigma_k - \mu_k}{\mu_j - \mu_k} \frac{[w_i, w_k][z_j, z_k]}{\omega_{kk}}.$$

To the third term apply the identity

$$[u, f][g, h] + [u, g][h, f] + [u, h][f, g] = 0,$$

which yields

$$\frac{[w_i, w_k][z_j, z_k]}{\omega_{kk}} = [w_i, z_j] - \frac{[w_i, z_k][w_k, z_j]}{\omega_{kk}}.$$

It follows that

$$[v_i(p_k), z_j(p_k)] = \frac{\mu_j - \sigma_k}{\mu_j - \mu_k}[w_i, z_j] + \frac{\sigma_k - \mu_k}{\mu_j - \mu_k}\frac{\sigma_i - \mu_j}{\sigma_i - \mu_k}\frac{[w_i, z_k][w_k, z_j]}{\omega_{kk}}.$$

It remains to multiply both sides of this equation by

$$\frac{\sigma_i - \mu_i}{\sigma_i - \mu_j}\frac{\mu_k - \sigma_i}{\sigma_k - \sigma_i}.$$

On the left hand side we obtain

$$\frac{\sigma_i - \mu_i}{\sigma_i - \mu_j}[w_i, (p_k), z_j(p_k)] = \omega_{ij}(p_k).$$

Using corresponding identities on the right hand side we arrive at

$$\omega_{ij}(p_k) = \frac{\mu_k - \sigma_i}{\sigma_k - \sigma_i}\frac{\sigma_k - \mu_j}{\mu_k - \mu_j}\left(\omega_{ij} - \frac{\omega_{ik}\omega_{kj}}{\omega_{kk}}\right)$$

as was to be proven.   ■

G Numerical Calculations

In this appendix the FORTRAN programs are described and listed that were used to produce the figures in Chapters 5 and 6. Since the actual plotting of these figures depends on the machine and software used, only those sub-routines are given that produce the data to be plotted. To make them easy to read, virtually no attempts was made to optimize them with respect to speed or code.

There are three subroutines KAPPA, MUE and EXPON. In each case, the output of the subroutine is a "function on [0, 1]" given by a vector *result* of length *len* + 3, where the value of the function at the point

$$x_k = \frac{k}{len}, \qquad 0 \le K \le len$$

is stored in the k + 2nd element of this vector. Its first and last element are only used during intermediate calculations and do not contain meaningful data.

KAPPA determines the function $q = \phi'(0, V_n)$, when the parameter j is zero, or its normalized eigenfunction g_j, when $j \ge 1$, according to the formulas of Theorem 5.1. Similarly, MUE determines the function $q = \phi'(0, W_n)$, when $j = 0$, or its normalized eigenfunction g_j, when $j \ge 1$.

This program makes use of the fact that q is always even, while g_j is even when j is odd and odd when j is even.

The syntax is

$$\text{CALL KAPPA}(result, j, n, t, vector, len)$$

and

$$\text{CALL MUE}(result, j, n, s, vector, len).$$

Here, *vector* is a vector of length *len* + 3 used to store intermediate results, and $-1 < s < 1$ is a rescaled time variable such that

$$t = (n + s)^2 \pi^2 - n^2 \pi^2.$$

See the program listings for the type of each parameter.

EXPON either determines $q = \exp_0(V_s)$ or $q = \mu_E^{-1}(s)$ for arguments $s = (s_1, s_2, \ldots)$ having only finitely many coordinates

$$s_{n_1}, \ldots, s_{n_m}$$

different from zero. In this case, the formulas of Theorem 5.3 and 6.2 are both of the form

$$q = -2 \frac{d^2}{dx^2} \log \det E,$$

where E is an $m \times m$-matrix with entries $\theta_{n_i n_j}(x, s_{n_i}, 0)$ and $\omega_{n_i n_j}(x, s_{n_i})$ respectively. For the sake of simplicity, only the case $1 \le m \le 3$ is supported.

The syntax is

$$\text{CALL EXPON}(result, nvec, svec, subname, vector, array, m, len).$$

Here, *nvec* and *svec* are vectors of length m containing the indices $n_l, \ldots, n_m$ and the arguments $s_{n_l}, \ldots, s_{n_m}$ respectively, *subname* is either THETA or OMEGA, the name of the subroutine calculating the entries of E, and *vector* and *array* are an auxiliary vector of length *len* + 3 and an auxiliary array of size $m \times m \times (len + 3)$ respectively. In case *subname* is OMEGA, the arguments in *svec* must lie in the interval $(-1, 1)$ and are rescaled to $(n_i + s_{n_i})^2 \pi^2 - n_i^2 \pi^2$, $1 \le i \le m$. See the program listing for the type of each parameter.

KAPPA, MUE and EXPON call a number of other subroutines, which we only describe briefly.

OMEGA calculates the ω-function at 0 using the identity

$$\omega_{nm}(x, \lambda) = \frac{t}{\lambda - \mu_m} + t \int_0^x w_n(s, \lambda) z_m(s) \, ds,$$

where $t = \lambda - \mu_n$. The program makes use of the fact that this function is even.

LOGDER approximates the second logarithmic derivative

$$-2\frac{d^2}{dx^2}\log u = -2\left(\frac{u''}{u} - \left(\frac{u'}{u}\right)^2\right)$$

of a function u by the second difference expression

$$\frac{2}{h^2 u_k^2}\left((u_{k-1} - 2u_k + u_{k+1})u_k - \frac{1}{4}(u_{k+1} - u_{k-1})^2\right).$$

Here, u_k is the value of u at the kth point of a grid of width h.

NORMAL normalizes a vector so that it has euclidian length 1.

DET yields the determinant of an $m \times m$-matrix. For the sake of simplicity, only the case $1 \le m \le 3$ is supported.

The calculation of the function w_n is speeded up by writing

$$w_n(x, \lambda, 0) = \cos\sqrt{\lambda}x + c_n(\lambda)\sin\sqrt{\lambda}x$$

with

$$c_n(\lambda) = \frac{(-1)^n - \cos\sqrt{\lambda}}{\sin\sqrt{\lambda}}.$$

The function COEFF yields the value of $c_n(\lambda)$ for given n and $\sqrt{\lambda}$. This value is passed as a parameter to the function W calculating w_n.

These programs were run on an IBM Personal Computer equipped with an 8087 math coprocessor. The plots were produced on a Hewlett Packard 7475 Plotter.

```
C          SUBROUTINE KAPPA(V,J,N,T,TH,L)
C          REAL*8 V(*),TH(*)
C          CALL THETA(TH,N,N,T,L)
           IF (J.EQ.0) THEN
C             CALL LOGDER(TH,V,L)
           ELSE IF (J.EQ.N) THEN
              FCT = EXP(T/2.)
              DO 1 K=2,L+2
                 X    = REAL(K-2)/L
                 V(K) = FCT*G(N,X)/TH(K)
C    1         CONTINUE
           ELSE
              CALL THETA(V,J,N,T,L)
              DO 2 K=2,L+2
                 X    = REAL(K-2)/L
                 V(K) = G(J,X)-G(N,X)*V(K)/TH(K)
```

```
      2     CONTINUE
            END IF
C
            RETURN
            END


            SUBROUTINE MUE(V,J,N,S,OM,L)
C
            REAL*8  SUM,V(*),OM(*)
            REAL    LAM,MU
C
            CALL OMEGA(OM,N,N,S,L)
C
            IF (J.EQ.0) THEN
              CALL LOGDER(OM,V,L)
C
            ELSE
              IF (J.EQ.N) THEN
                DO 1 K=2, (L+4)/2
                    X    = REAL(K-2)/L
                    V(K) = Z(N,X)/OM(K)
      1         CONTINUE
              ELSE
                PI   = 3.141593
                RL   = PI*(N+S)
                LAM  = RL**2
                T    = LAM-MU(N)
                CO   = COEFF(N,RL)
                H    = 1./L
                SUM  = 0.
C
                DO 2 K=2, (L+4)/2
                    X    = REAL(K-2)/L
                    SUM  = SUM+Z(J,X)*Z(N,X)
                    V(K) = Z(J,X)-T*W(CO,RL,X)*SUM*H/OM(K)
      2         CONTINUE
              END IF
C
              SIG = 2*MOD(J,2)-1
              DO 3 K=(L+4)/2,L+2
                  V(K) = SIG*V(L+4-K)
      3       CONTINUE
C
              CALL NORMAL(V,L)
            END IF
C
            RETURN
            END


            SUBROUTINE EXPON(V,N,S,ELEM,W,MAT,M,L)
C
            REAL*8      V(*),W(*),MAT(M,M,*),DET
            REAL        S(*)
            INTEGER     N(*)
            EXTERNAL    ELEM
C
            DO 1 I=1,M
```

```
              DO 1 J = 1,M
                 CALL ELEM(W,N(I),N(J),S(I),L)
                 DO 2 K = 1,L + 3
                    MAT(I,J,K) = W(K)
        2        CONTINUE
        1     CONTINUE
C
              DO 3 K = 1,L + 3
                 W(K) = DET(MAT(1,1,K),M)
        3     CONTINUE
C
C             CALL LOGDER(W,V,L)
C
              RETURN
              END




              SUBROUTINE THETA(TH,N,M,T,L)
C
C             REAL*8 TH(*),SUM
              FCT = EXP(T) - 1.
              H   = 1./L
              X   = 1. + H/2.
              SUM = - G(N,X)*G(M,X)
C
              IF (N.EQ.M) THEN
                 DO 1 K = L + 3,1, - 1
                    TH(K) = 1. + FCT*SUM*H
                    SUM   = SUM + G(N,X)**2
                    X     = X - H
        1        CONTINUE
              ELSE
                 DO 2 K = L + 3,1, - 1
                    TH(K) = FCT*SUM*H
                    SUM   = SUM + G(N,X)*G(M,X)
                    X     = X - H
        2        CONTINUE
              END IF
C
              RETURN
              END




              SUBROUTINE OMEGA(OM,N,M,S,L)
C
C             REAL*8 SUM,OM(*)
              REAL    LAM,MU
C
              PI  = 3.141593
              RL  = PI*(N + S)
              LAM = RL**2
              T   = LAM - MU(N)
              CO  = COEFF(N,RL)
C
              IF (N.EQ.M) THEN
                 TRM = 1.
```

```
         ELSE
           TRM = T/(LAM – MU(M))
         END IF
C
         H     = 1./L
         X     = – H/2.
C        SUM = – W(CO,RL,X)*Z(M,X)
         DO  1 K = 1,(L + 4)/2
            OM(K) = TRM + T*SUM*H
            SUM    = SUM + W(CO,RL,X)*Z(M,X)
            X      = X + H
C     1  CONTINUE
         DO  2 K = (L + 4)/2,L + 3
            OM(K) = OM(L + 4 – K)
C     2  CONTINUE
C        RETURN
         END



C        SUBROUTINE LOGDER(U,V,L)
C        REAL*8 UM,UN,UP,U(*),V(*)
C        FCT = – 2*(L**2)
         DO  1 K = 2,L + 2
            UM = U(K – 1)
            UN = U(K)
            UP = U(K + 1)
            V(K) = FCT*((UP – 2*UN + UM)*UN –
                   ((UP – UM)/2)**2)/UN**2
C     1  CONTINUE
         RETURN
         END



C        REAL*8 FUNCTION DET(A,M)
C        REAL*8 A(M,M)
         IF (M.EQ.1) THEN
C           DET = A(1,1)
         ELSE IF (M.EQ.2) THEN
            DET = A(1,1)*A(2,2) – A(1,2)*A(2,1)
C        ELSE IF (M.EQ.3) THEN
            DET =    A(1,1)*(A(2,2)*A(3,3) – A(2,3)*A(3,2))
      *           – A(1,2)*(A(2,1)*A(3,3) – A(2,3)*A(3,1))
      *           + A(1,3)*(A(2,1)*A(3,2) – A(2,2)*A(3,1))
C        ELSE
C           M > 3 not supported
         END IF
C        END
```

```
C          SUBROUTINE NORMAL(V,L)
C          REAL*8 V(*),SUM,DSQRT
C          SUM = 0.
           DO 1 K=2,L+2
             SUM = SUM+V(K)**2
C     1    CONTINUE
C          FCT = DSQRT(L/SUM)
           DO 2 K=2,L+2
             V(K) = FCT*V(K)
C     2    CONTINUE
           RETURN
           END


C          REAL FUNCTION MU(N)
           PI  = 3.141593
C          MU = (PI*N)**2
           END


C          REAL FUNCTION G(N,X)
C          DATA PI,ROOT /3.141593,1.414214/
C          G = ROOT*SIN(PI*N*X)
           END


C          REAL FUNCTION Z(N,X)
           PI  = 3.141593
           PIN = PI*N
           Z   = SIN(PIN*X)/PIN
C          END


C          REAL FUNCTION W(CO,RL,X)
           RLX = RL*X
           W   = COS(RLX)+CO*SIN(RLX)
C          END


C          REAL FUNCTION COEFF(N,RL)
C          DATA PI,EPS /3.141593,0.01/
           IF (ABS(RL-PI*N).GT.EPS) THEN
             COEFF = (1-2*MOD(N,2)-COS(RL))/SIN(RL)
           ELSE
             COEFF = SIN(RL)/COS(RL)
           END IF
C          END
```

References

[Ah] L. Ahlfors, *Complex Analysis*. McGraw-Hill, New York, 1979.

[Am] V. Ambarzumian, "Über eine Frage der Eigenwerttheorie". *Zeitschr. f. Phys.* **53**, 690–695, 1929.

[Ba] A. Balakrishnan, *Applied Functional Analysis*. Springer, New York, 1981.

[Bo] G. Borg, "Eine Umkehrung der Sturm-Liouvilleschen Eigenwertaufgabe". *Acta Math.* **78**, 1–96, 1946.

[BR] G. Birkhoff, G. Rota, "On the completeness of Sturm-Liouville expansions". *Am. Math. Monthly* **67**, 835–851, 1960.

[CS] K. Chadan, P. Sabatier, *Inverse Problems in Quantum Scattering Theory*. Springer, New York, 1977.

[Da1] G. Darboux, "Sur la représentations sphérique des surfaces". *Compt. Rend.* **94**, 1343–1345, 1882.

[Da2] G. Darboux, *Leçons sur le Théorie générale des Surfaces et les applications géométrique du calcul infinitesimal. Vol. 2*, Gauthier Villars, Paris, 1915.

[DaT] B. Dahlberg, E. Trubowitz, "The inverse Sturm-Liouville problem III". *Comm. Pure Appl. Math.* **37**, 255–267, 1984.

[De] P. Deift, "Applications of a commutation formula". *Duke Math. J.* **45**, 267–310, 1978.

[DeT] P. Deift, E. Trubowitz, "Inverse scattering on the line". *Comm. Pure Appl. Math.* **32**, 121–251, 1979.

[Di] J. Dieudonné, *Foundations of Modern Analysis*. Academic Press, New York, 1969.

[Din] S. Dineen, *Complex Analysis in Locally Convex Spaces*. North Holland, Amsterdam, 1981.

[GL] I. Gel'fand, B. Levitan, "On the determination of a differential equation from its spectral function". *Amer. Math. Soc. Transl.* **1**(2), 253–304, 1955. Russian: *Izv. Akad. Nauk SSSR* **15**, 309–360, 1951.

[HL]   H. Hochstadt, B. Lieberman, "An inverse Sturm-Liouville problem with mixed given data". *SIAM J. Appl. Math.* **34**, 676–680, 1978.

[IMT]   E. Isaacson, H. McKean, E. Trubowitz, "The inverse Sturm-Liouville problem II". *Comm. Pure Appl. Math.* **37**, 1–11, 1984.

[IT]   E. Isaacson, E. Trubowitz, "The inverse Sturm-Liouville problem I". *Comm. Pure Appl. Math.* **36**, 767–783, 1983.

[JK]   R. Jost, W. Kohn, "Equivalent Potentials". *Phys. Rev. Ser. 2* **88**, 382–385, 1952.

[La1]   S. Lang, *Real Analysis.* Addison-Wesley, Reading, 1969.

[La2]   S. Lang, *Differentiable Manifolds.* Addison-Wesley, Reading, 1972.

[Le]   N. Levinson, "The inverse Sturm-Liouville problem". *Mat. Tidsskr. B.*, 25–30, 1949.

[Li]   J. Liouville, "Mémoire sur le développement des fonctions ou parites de fonctions en séries dont les divers termes sont assujettis à satisfaire à une même équation differentielles du second ordre contenant un paramètre variable". *J. de Math.* **1**, 253–265, 1836; **2**, 16–35 and 418–436, 1837.

[MT]   H. McKean, E. Trubowitz, "Hill's operator and hyperelliptic function theory in the presence of infinitely many branch points". *Comm. Pure Appl. Math.* **29**, 143–226, 1976.

[Na]   L. Nachbin, "Topology on Spaces of Holomorphic Mappings". *Ergebnisse der Mathematik und ihrer Grenzgebiete* **47**. Springer, 1969.

[RS]   M. Reed, B. Simon, *Methods of Modern Mathematical Physics, volume I and IV.* Academic Press, New York, 1980 and 1978.

[Ru]   W. Rudin, *Functional Analysis.* McGraw-Hill, New York, 1973.

[St]   C. Sturm, "Sur les équations differentielle linéares du second ordre". *J. de Math.* **1**, 106–186, 1836.

[Ti]   E. Titchmarsh, *The Theory of Functions.* Clarendon Press, Oxford, 1932.

[Wa]   W. Wasow, *Asymptotic Expansions for Ordinary Differential Equations.* Interscience, New York, 1965.

[We]   H. Weyl, "Über gewöhnliche Differentialgleichungen mit Singularitäten und die zugehörigen Entwicklungen willkürlicher Funktionen". *Math. Ann.* **68**, 220–269, 1910.

Index

187

Index of Notations

The symbols are listed in their order of appearance.

PURE AND APPLIED MATHEMATICS

H. Bass, A. Borel, J. Moser, and S.-T. Yau, editors

PURE AND APPLIED MATHEMATICS

* Presently out of print